ISBN 978-0-266-84867-7
PIBN 10898740

This book is a reproduction of an important historical work. Forgotten Books uses
state-of-the-art technology to digitally reconstruct the work, preserving the original format
whilst repairing imperfections present in the aged copy. In rare cases, an imperfection in
the original, such as a blemish or missing page, may be replicated in our edition. We do,
however, repair the vast majority of imperfections successfully; any imperfections that
remain are intentionally left to preserve the state of such historical works.

MINISTRY OF PUBLIC WORKS, EGYPT.

ZOOLOGICAL SERVICE.

REPORT

ON A

ZOOLOGICAL MISSION TO INDIA

IN 1913.

BY

Capt. S. S. FLOWER,

DIRECTOR, EGYPTIAN GOVERNMENT ZOOLOGICAL SERVICE.

(Publication No. 26.)

42080

CAIRO.
GOVERNMENT PRESS.

To be obtained, either directly or through any Bookseller,
from the PUBLICATIONS OFFICE, Government Press, Bulâq ; or from the SALE-ROOM,
Geological Museum, Ministry of Public Works Gardens.

1914.

PRICE P.T. 24 (or Five Shillings).

CONTENTS

LIST OF PLATES.

Plates IX, X, XI, and XII are from photographs kindly taken for the purpose by Mr. S. A. Christopher of Rangoon; Plate VI is from a photograph placed at our disposal by the authorities of the Mysore Zoological Gardens; the remaining plates are from photographs by Captain Flower.

PREFACE.

In every zoological garden visited in India there were particular features of interest, and in each there were new facts of menagerie technique to be learnt.

Three zoological gardens proved to be of the greatest importance: Calcutta, for its large collections; Trivandrum, thanks to the work of Mr. H. S. Ferguson, for being such a complete and scientifically arranged institution; Peshawar, for the admirable way in which all the animals were looked after.

I have to express my deep sense of obligation to Field Marshal Viscount Kitchener of Khartoum for the invaluable advice he gave me as to what places I ought to visit in India and for kindly giving me introductions which enabled me to meet important officials, to make pleasant friends, and to employ the time at my disposal to the best advantage. I wish also to be allowed to thank H.H. the Maharaja of Mysore, H.H. the Maharaja of Trivandrum, and H.H. the Maharaja of Alwar for the courtesies and assistance I received while in their respective dominions. To the officials of all the institutions visited my thanks are due for their kindness in showing me the collections under their charge and in giving me the information I sought for, and I am particularly indebted to the officers of the Bombay Natural History Society for many acts of consideration. My visit to Calcutta was very interesting and the time profitably spent, thanks to the kindness of Dr. E. P. Harrison. Finally, I have to thank Mr. A. H. Froom and the Peninsular and Oriental Steam Navigation Company for granting a free passage from India to Egypt for the two young crocodiles which the officials of the Karachi Municipal Gardens were so good as to entrust me with for conveyance to the Egyptian Government Gardens at Giza.

The scientific nomenclature used in this Report is as far as possible that of the invaluable volumes on the Fauna of British India edited by the late Mr. W. T. Blanford.

REPORT ON ZOOLOGICAL MISSION TO INDIA IN 1913.

I.—OBJECTS OF MISSION.

(1) With a view to making improvements in similar matters in the Egyptian Zoological Gardens, to specially inquire into the following subjects :—

    (i) Methods of managing and housing the larger and more dangerous animals such as lions, elephants, rhinoceros, buffaloes, crocodiles, and pythons.

    (ii) Methods of obtaining, storing, distributing, and checking forage, material, etc.

    (iii) Distribution of responsibilities and work amongst the staff.

    (iv) Methods of keeping accounts and checking expenditure.

    (v) Methods of registering the collection.

    (vi) Methods of managing large crowds of visitors, and of keeping order among them.

(2) With a view to disposing of such surplus animals as there may be from time to time in the Egyptian collections, and also to helping the Indian Zoological Gardens by supplying them with the beasts or birds that they want, to ascertain the *desiderata* of the various Indian institutions and the terms (by purchase or exchange) that might be expected.

Incidentally, as far as time permitted, I looked at every animal in each zoological garden, to note its condition, food, accommodation, etc., as compared to our menagerie at Giza.

The reason for visiting stables and kennels (*see* Chap. III, p. 5) is : that of all the various species of animals that mankind has kept alive in conditions differing from those of nature there are two species—the horse and the dog—on the care of which the human intellect has been specially exercised from century to century for untold thousands of years, and therefore the director of a zoological garden should be always on the alert to learn from the experiences of the horse-master and the kennel-man facts that may prove useful in the management of less known animals.

II.—ITINERARY.

1913.

			Miles.
April	1	Cairo to Port Said.	
	2–11	Port Said, *viâ* Aden, to Bombay ...	3,518
	12–14	Bombay, *viâ* Guntakal, to Bangalore	
	16	Bangalore to Mysore	
	18	Mysore to Bangalore	1,483
	19–21	Bangalore, *viâ* Erode and Maniyachi, to Quilon	
	22–23	Quilon to Trivandrum (by boat)	
	27	Trivandrum to Quilon (by motor car)	592
	27–29	Quilon, *viâ* Maniyachi, to Madras...	
May	2– 8	Madras, *viâ* Rangoon, to Calcutta...	1,994
	14–16	Calcutta, *viâ* Agra, to Jaipur... ...	
	17	Jaipur to Alwar	1,670
	19–20	Alwar, *viâ* Bhatinda, to Lahore ...	
	21–22	Lahore to Peshawar	
	23–25	Peshawar, *viâ* Lahore, to Karachi...	
	28–29	Karachi, *viâ* Luni, to Ahmedabad...	2,064
	31	Ahmedabad to Baroda	
June	1	Baroda to Bombay	
	7–18	Bombay, *viâ* Aden, to Port Said ...	3,518
	18	Port Said to Cairo.	

Total distance travelled from Port Said back to Port Said was 14,839 miles, or 23,880 kilometres.

Total distance travelled from Bombay back to Bombay was 7,803 miles, or 12,557 kilometres.

III.—LIST OF PLACES OF ZOOLOGICAL AND BOTANICAL INTEREST IN INDIA, VISITED APRIL-JUNE, 1913.

LIST OF PLACES VISITED (*continued*).

Some small gardens, tanks, and collections of heads and horns of game animals were also visited in Calcutta, Karachi, Madras, Trivandrum, etc.

IV.—LIST OF INDIAN ZOOLOGICAL GARDENS.

	Town.	Province or State.	Date when Started.	Proprietor.	Official in Charge.	Area of Land, Occupied or Available.			Usual Admission.	Name of Garden.
						Acres.	Hectares.	Remarks.		
1	Bangalore.	Mysore.	c. 1855	Government.	Mr. G. H. Krumbiegel, Supt. ...	100	40·46	—	Free.	Lal Bagh.
2	Baroda.	Baroda.	?	The Maharaja.	?	?	?	Apparently over 50 acres.	Free.	Public Gardens.
3	Bombay.	Bombay.	c. 1870	Municipality.	Mr. C. D. Mahaluxmivala, Supt....	50	20·23	Appears to be less ...	Free.	Victoria Gardens.
4	Calcutta.	Bengal.	1875	Government.	Mr. B. K. Basu, Supt.	64	25·89	—	1 Anna.	Zoological Gardens.
5	Jaipur.	Jaipur.	? 1875	The Maharaja.	?	36	14·56	Appears to be more ...	Free.	Ram Newas Gardens.
6	Karachi.	Sind.	1881	Municipality.	Mr. Ali Murad, Supt.	46	18·61	—	Free.	Zoological Gardens.
7	Lahore.	Punjaub.	?	Government.	Mr. A. W. Pinto, Curator ...	112	45·32	These are the areas of the whole gardens; the menageries are concentraced, in each case, in one part only.	Free.	Lawrence Gardens.
8	Madras.	Madras.	Before 1858	Municipality.	Mr. H. Garwood, Supt.	116	46·94		½ Anna.	People's Park.
9	Mysore.	Mysore.	1892	The Maharaja.	Mr. A. C. Hughes, Supt. ...	35	14·16	—	1 Anna.	Sri Chamarajendra Zoological Gardens.
10	Peshawar.	North-West Frontier.	c. 1909	Municipality.	Capt. J. G. L. Ranking, Hon. Dir.	?	?	Apparently over 50 acres.	Free.	Shahi Bagh.
11	Rangoon.	Burma.	1906	Government.	Dr. R. M. Sen, Supt.	80	32·37	—	1 Anna.	Victoria Memorial Park.
12	Trivandrum.	Travancore.	1859	Government.	Mr. A. J. Vieyra, Director. ...	50½	20·43	—	Free.	Public Gardens.

Besides the zoological gardens shown on the opposite page, there are others open to the public at *Burdwan*, in Bengal (belonging to the Maharaja of Burdwan), at *Nagpore*, in the Central Provinces, and at *Secunderabad*, Hyderabad, in the Deccan (belonging to the Nizam). There is, or was, also a small one at *Jubbulpore*, in the Central Provinces. The Amir of Afghanistan maintains a collection of wild animals at *Jellalabad*, and several of the Maharajas have private menageries or a few animals in their public gardens, as at Alwar, Gwalior, Indore, and Kolhapur.

V.—ANALYSIS OF THE COLLECTIONS OF LIVE ANIMALS SEEN

	1 BANGALORE		2 BARODA		3 BOMBAY		4 CALCUTTA		5 JAIPUR		
	Specimens.	Forms.	Specimens.	Forms.	Specimens.	Forms.	Specimens.	Forms.	Specimens.	Forms.	Specimens.
I.—MAMMALS.											
1. Mias, or Orang-Utan	1	1	—	—	1	—	3	1	—	—	—
2. Gibbon, *Hylobates*	—	—	—	—	1	1	3	2	—	—	—
3. Old World Monkeys, *Cercopithecidae.*	3	3	20	9	c. 45*	14	58	23	c. 22	8	24
4. American Monkeys, *Cebidae*	—	—	—	—	1	1	4	4	1	1	—
5. Marmosets, *Hapalidae*	—	—	1	1	1	—	c. 6	3	2	1	—
6. Lemurs	5	3	12	5	6	4	17	6	3	3	1
Total Primates	9	7	33	15	c. 53	20	c. 91	39	c. 28	13	25
7. Lions	1	1	2	1	4	1	4	1	—	—	3
8. Tigers	2	1	—	—	3	1	8	1	7†	1	2
9. Leopards	6	1	5	1	6	1	7	4	4	2	12
10. Other Cats of the *Felidae*	—	—	—	—	2	2	12	7	—	—	2
11. Civets, Mongeese, etc., *Viverridae*	—	—	—	—	6	3	5	4	1	1	2
12. Hyaenas	1	1	1	1	2	1	3	1	—	—	3
13. Wolves, Jackals, and Foxes	5	1	2	1	16	5	8	4	4	3	5
14. Otters, Badgers, etc., *Mustelidae*	—	—	—	—	4	2	6	3	1	1	2
15. Racoons, Pandas, etc., *Procyonidae*	—	—	—	—	2	1	7	3	—	—	—
16. Bears, *Ursidae*	4	2	2	1	12	3	8	2	4	2	4
Total Carnivores	19	7	12	5	57	20	68	30	21	10	35
17. Porcupines, Squirrels, etc., *Rodentia.* §	?	2	?	2	5	1	c. 24	14	6	5	?
18. Elephants	—	—	—	—	1	1	1	1	—	—	—
19. Rhinoceros	—	—	—	—	—	—	2	1	—	—	—
20. Zebras, etc., *Equidae*	—	—	—	—	2	1	4	2	4	1	1
21. Tapirs	—	—	1	1	1	1	4	1	—	—	—
22. Cattle	—	—	—	—	—	—	9	5	—	—	—
23. Antelopes	5	2	c. 14	4	c. 35	8	27	8	c. 15	1	23
24. Goat-Antelopes	—	—	—	—	—	—	—	—	—	—	1
25. Goats	—	—	—	—	—	—	1	1	—	—	5
26. Sheep	—	—	—	—	1	1	11	1	—	—	7
27. Giraffes	—	—	—	—	—	—	—	—	—	—	—
28. Deer, *Cervidae*	c. 30	3	c. 20	5	14	3	58	8	c. 15	2	35
29. Mouse-Deer, *Tragulidae*	—	—	—	—	1	1	5	2	1	1	1
30. Camels and Lamas	—	—	—	—	6	2	c. 8	2	—	—	—
31. Hippopotamus	—	—	—	—	—	—	2	1	—	—	—
32. Pigs, *Suidae*	—	—	—	—	3	1	8	1	—	—	9
Total Ungulates	c. 35	5	c. 35	10	c. 64	19	c. 140	34	c. 35	5	82
33. Ant-eaters, etc., *Edentata*	—	—	—	—	—	—	—	—	—	—	—
34. Kangaroos, etc., *Marsupialia*	—	—	c. 8	1	5	2	c. 16	9	1	1	2
TOTAL MAMMALS	c. 63	21	c. 90	33	c. 184	62	c. 339	126	c. 91	34	145

* " *c.*" (*circa*) signifies that the number is approximate.
† JAIPUR : 5 tigers and 2 leopards in the city are included.
‡ MYSORE : 2 leopards in the city are included.

TWELVE ZOOLOGICAL GARDENS : APRIL 12 TO JUNE 6, 191[...]

	7 LAHORE Specimens	7 LAHORE Forms	8 MADRAS Specimens	8 MADRAS Forms	9 MYSORE Specimens	9 MYSORE Forms	10 PESHAWAR Specimens	10 PESHAWAR Forms	11 RANGOON Specimens	11 RANGOON Forms	12 TRIVANDRUM Specimens	12 TRIVANDRUM Forms	Number of Individual Anim[...]
	—	—	1	1	2	1	—	—	—	—	1	1	8 Mias, *Simia satyrus*.
	—	—	1	—	1	1	—	—	2	2	—	—	7 Gibbons.
	11	5	25	6	19	12	8	4	35	11	15	8	c. 285 Old World Monkeys.
	—	—	—	—	4	3	—	—	—	—	1	1	11 American Monkeys.
	—	—	—	—	—	—	1	1	—	—	—	—	c. 10 Marmosets.
	2	1	2	1	7	3	1	1	3	1	2	2	61 Lemurs and Lorises.
	13	6	28	8	33	20	10	6	40	14	19	12	
	—	—	1	1	2	1	2	1	2	1	3	1	24 Lions.
	—	—	2	1	2	1	2	1	—	—	3	1	34 Tigers : including 3 seen
	4	1	6	1	6‡	2	2	1	8	1	7	2	75 Leopards : including 2 a
	—	—	—	—	—	—	—	—	8	2	3	2	27 Jungle Cats, Lynxes, Ch
	1	1	9	3	4	3	—	—	9	4	11	7	48 Civet Cats, Mongeese, B etc.
	1	1	4	1	2	1	1	1	2	1	4	1	24 Striped Hyaenas.
	2	2	—	—	9	4	2	2	4	1	2	1	59 *Canidae*
	4	1	—	—	—	—	2	2	4	2	4	3	27 Otters, Badgers, etc.
	—	—	—	—	2	1	—	—	1	1	—	—	12 Pandas, Racoons, and Co
	5	3	5	3	6	3	4	3	6	2	3	1	66 Bears : including 3 at Al
	17	9	27	10	33	16	15	11	44	15	40	19	
	?	2	?	2	?	1	9	5	c. 17	10	15	5	
	—	—	1	1	2	1	—	—	2	1	1	1	53 Elephants. ¶
	—	—	1	1	—	—	—	—	1	1	—	—	4 Rhinoceros.
	3	2	—	—	4	2	2	2	1	1	1	1	22 Donkeys, Zebras, etc.
	—	—	1	1	—	—	—	—	3	1	—	—	10 Malay Tapirs.
	—	—	—	—	—	—	—	—	10	4	2	2	21 Cattle.
	11	2	7	5	25	4	8	2	7	2	5	2	c. 185 Antelopes : including 3 a abad.
	—	—	—	—	—	—	2	1	—	—	—	—	3 Goral.
	1	1	—	—	2	2	2	2	—	—	—	—	11 Goats.
	4	1	1	1	2	2	—	—	—	—	—	—	26 Sheep.
	—	—	—	—	3	1	—	—	—	—	—	—	3 Giraffes.
	c. 8	2	11	2	51	3	5	2	47	5	14	4	c. 308 Deer.
	—	—	—	—	1	1	—	—	—	—	—	—	9 Mouse-Deer.
	—	—	3	2	13	4	—	—	1	1	3	2	34 Camels and Lamas.
	—	—	—	—	—	—	—	—	—	—	—	—	2 Hippopotamus.
	—	—	—	—	4	1	—	—	4	2	2	1	30 Wild Swine.
	c. 27	8	25	13	107	21	19	9	76	18	28	13	
	1	1	3	2	15	2	1	1	7	2	2	1	
	c. 58	26	83	35	188	60	54	32	c. 184	59	105	52	Approximately 1,638 individ mals noted.

§ *Rodentia* : the numbers of specimens of domestic rabbits and guinea-pigs are not included.
¶ Elephants : this includes 5 other elephants seen at Rangoon, 23 at Alwar, 14 at Trivandrum, and 3 at Jaip[...] elephants seen at Mysore are not included.

YSIS OF THE COLLECTIONS OF LIVE ANIMALS SEEN IN

	1 BANGALORE.		2 BARODA.		3 BOMBAY.		4 CALCUTTA.		5 JAIPUR.		6 KARACHI.	
	Specimens.	Forms.	Specimens.	Forms.	Specimens.	Forms.	Specimens.	Forms.	Specimens.	Forms.	Specimens.	Forms.
Total	c. 63	21	c. 90	33	c. 184	62	c. 339	126	c. 91	34	145	44
ches, etc., *Passeres* ...	—	—	c. 3	2	?	14	?	c. 35	?	c. 50	?	9
ıckoos, etc., *Picariae*...	—	—	—	—	c. 7	2	?	c. 9	c. 8	3	?	1
tacı...	—	—	c. 25	14	c. 40	15	?	e. 27	c. 123	c. 26	?	6
!	—	—	—	—	—	—	6	2	—	—	2	1
y, *Accipitres*...	—	—	1	1	—	—	c. 19	c. 13	—	—	1	1
norants,etc.,*Steganopo-*												
... ...	—	—	1	1	c. 6	1	?	3	c. 22	4	c. 7	4
ks, etc., *Herodiones* ...	—	—	11	5	?	6	?	14	39	7	34	8
Ducks, etc., *Anseres* ...	12	2	c. 20	8	?	17	?	34	c. 200	18	?	18
umbae *	—	—	c. 12	c. 5	?	3	?	c. 13	?	c. 6	?	3
, *Pterocletes*	—	—	—	—	—	—	?	3	—	—	—	—
quail, etc., *Gallinae* † ...	2	2	c. 12	5	?	15	?	23	?	14	?	8
etc , *Fulicariae*	—	—	1	1	?	1	?	7	c. 20	2	?	2
Alectorides	—	—	1	1	?	4	?	7	c. 11	3	c. 3	2
, *Limicolae*	—	—	—	—	—	—	2	1	c. 32	3	—	—
ıe	—	—	—	—	?	c. 3	—	—	—	—	—	—
Casuarius	—	—	—	—	—	—	—	—	—	—	1	1
aeus	2	1	—	—	1	1	3	1	2	1	—	—
iches, *Struthio*	—	—	3	1	3	1	3	1	2	1	—	—
striches, *Rhea*	—	—	1	1	8	1	1	1	—	—	—	—
	16	5	c. 91	c. 35	?	c. 84	?	c. 194	?	c. 138	?	64
S.												
ıelonia	—	—	3	1	9	2	c. 25	10	3	1	3	1
Crocodilia	—	—	2	1	1	1	21	4	—	—	3	1
ertilia	—	—	—	—	—	—	c. 17	7	1	1	—	—
Rhiptoglossa	—	—	—	—	—	—	2	1	—	—	—	—
ıdia...	—	—	—	—	—	—	c. 40	c. 19	—	—	—	—
	—	—	5	2	10	3	c. 105	c. 41	4	2	6	2
	—	—	—	—	—	—	?	1	—	—	—	—
	c. 79	26	c. 186	c. 70	?	c. 149	?	c. 362	?	c. 174	?	110

*: the numbers of specimens of domestic pigeons are not included.
: " " " " " " poultry " " "

SONS OF THE COLLECTIONS.—From the above tables and from my

ımber of Different ns of Mammals.	B.—Number of Individual Mammals.	C.—Number of Different Forms of Birds.	D.—Number of Individual Birds.
Calcutta.	Calcutta.	Calcutta.	Calcutta.
Bombay.	Mysore.	Jaipur.	Jaipur.
Mysore.	Bombay.	Bombay.	Lahore.
Rangoon.	Rangoon.	Lahore.	Bombay.
Trivandrum.	Karachi.	Peshawar.	Karachi.

he collections of live animals in the Bombay Natural History Society's Museum and in the Madras Museum and

TWELVE ZOOLOGICAL GARDENS: APRIL 12 TO JUNE 6, 1

7 LAHORE		8 MADRAS		9 MYSORE		10 PESHAWAR		11 RANGOON		12 TRIVANDRUM		Number of Individu...
Specimens	Forms	Specimens	Forms	Specimens	Forms	Specimens	Forms	Specimens	Forms	Specimens	Forms	
c. 58	26	83	35	188	60	54	32	c. 184	59	105	52	
?	c. 29	c. 23	c. 6	?	9	?	27	5	3	5	4	
				1	1	c. 3	2	1	1	15	6	
?	c. 16	12	6	c. 22	12	?	20	c. 17	10	10	6	
—	—	—	—	2	2	2	2	3	1	4	2	
								—	—	8	7	
1	1	2	1	2	2	—	—	1	1	1	1	
?	2	11	4	3	2	—	—	5	4	9	6	
?	12	6	3	?	16	?	7	c. 30	8	10	4	
?	5	c. 11	5	?	7	?	5	17	8	7	3	
?	1											
?	13	13	7	?	12	?	7	c. 70	10	c. 17	8	
								2	1			
2	1	1	1	5	2	6	3	4	3	1	1	
—	—	10	3	—	—	—	—	—	—	1	1	
—	—	1	1	—	—	—	—	2	2	3	1	7 Cassowaries.
—	—	2	1	2	1	—	—	c. 2	1	—	—	c. 14 Emus.
—	—	—	—	1	1	—	—	—	—	—	—	12 Ostriches.
—	—	—	—	1	1	—	—	—	—	—	—	11 Rheas.
?	c. 80	c. 92	c. 38	?	68	?	73	c. 159	53	c. 91	50	
.	—	—	—	1	1	—	—	c. 7	2	6	1	c. 57 Tortoises, not in the Jaipur
—	—	5	2	5	1	—	—	1	1	4	1	80 Crocodilians.
—	—	—	—	1	1	—	—	1	1	1	1	c. 21 Lizards.
—	—	—	—	3	1	—	—	—	—	—	—	2 Chameleons.
—	—	1	1	—	—	—	—	2	1	22	11	c. 68 Snakes.
—	—	6	3	10	4	—	—	c. 11	5	33	14	c. 228 individual Re...
?	1	—	—	—	—	—	—	—	—	c. 7	2	
?	c. 107	c. 181	c. 76	?	132	?	105	c. 354	117	c. 236	118	

‡ *Crocodilia*: this includes 9 other crocodiles seen at Jaipur, 27 at " Mugger Pir," and 2 at Kankeria.

notes on the different institutions the zoological gardens may be arranged

E.—Number of Different Forms of Reptiles.	F.—Number of Individual Reptiles.	G.—Numbers both of Forms and Individuals, of all Classes.
Calcutta. Trivandrum.	Calcutta. Trivandrum.	Calcutta. Jaipur. Bombay. Mysore. Rangoon.
—	—	
—	—	

Aquarium are not included in the above tables.

VI.—NOTES ON ELEPHANTS.

1. Elephants seen in India and Burma.
2. Height of Elephants.
3. Size of Elephants' Tusks.
4. Age of Elephants.
5. Elephants' Face-glands.
6. Colour of Elephants' Eyes.

1.—Elephants seen in India and Burma.

INDIA.

At Alwar I saw twenty-three elephants; two were big males (*see* Pl. I) and the rest females. The mahouts and coolies in the elephant lines are all Mohammedans.

At Trivandrum I saw fourteen elephants, eleven fine tuskers and three females; each elephant has from one to three mahouts, according to its temper, and of course coolies as well. The mahouts are all Hindoos, Nayers of Travancore. I am told that many are of good family and landowners; all are hereditary mahouts.

At Mysore and at Jaipur I was also fortunate enough to see some elephants; at both these places the mahouts were Mohammedans.

I am indebted to Mr. C. V. Subramania Raj Urs, Superintendent of the Palace Elephants, Mysore, to Mr. C. R. Parameswaren Pillay, Superintendent of the Royal Stables, Trivandrum, and to Mr. Fatai Naseeb Khan, Superintendent of the Fŷl-Khana, Alwar, for their kindness in showing me the elephants in their respective charges, and in giving me very much interesting and useful information about Indian elephants and their management.

BURMA.

At Rangoon on May 7, 1913, Mr. S. A. Christopher and I visited all the timber yards, from Monkey Point to upstream of Aloon, to find out how many elephants were actually employed in Rangoon.

The work of stacking wood, in which elephants were employed till a few years ago, is now more efficiently and economically performed by overhead machinery worked by electricity.

We saw five elephants in all. One was "must," and chained up in a shed: it had recently severely wounded a man. The other four elephants were employed in rough and heavy work. Logs of teak are floated at high tide into a creek, and at low water the elephants are taken into the mud and drag out the logs and place them where required on dry land.

In the Rangoon Zoological Gardens I saw two young Burmese elephants, a male about two years and seven months old, and a female about one year four months old; their general appearance was very different from the elephants that I had been seeing in Southern India, but it would be necessary to compare a large series of both these races of elephants before one could definitely state on what grounds it was necessary to constitute them as belonging to two different subspecies, and the races occurring in Siam, Sumatra,* etc., should also be considered. The five adult Burmese elephants mentioned above seemed chiefly to differ from the elephants of Mysore and Travancore in being much more hairy.

The mahouts that I saw in Rangoon were mostly Mohammedans from Chittagong.

2.—Height of Elephants.

The late Mr. W. T. Blanford, F.R.S. ("Fauna of British India," Mammalia, 1891, p. 463), writes:—

"The vertical height at the shoulder in adult elephants is almost exactly twice the circumference of the fore foot. Adult males do not as a rule exceed 9 feet, females 8 in height, but a male has been measured by Sanderson as much

---

* *Elephas indicus sumatranus.* A male from Lower Landak, East Sumatra, can be seen stuffed in the Museum at Munich, Bavaria. (Oct. 7, 1913. S. S. F.).

as 10 feet 7½ inches; Col. Hamilton says that Sir V. Brooke killed one of 11 feet."

The tallest male elephant that I actually saw measured in India was "Chandra Sekaren," belonging to H.H. the Maharaja of Travancore; this animal was at the shoulder 10 feet (3·04 metres).

The tallest female elephant that I actually saw measured in India was one belonging to H.H. the Maharaja of Alwar; she was at the shoulder just 9 feet (2·74 metres).

In the Indian Museum, Calcutta, I saw the skeleton of Mr. W. M. Smith's Bilkandi elephant, January 19, 1870. Mr. Blanford wrote (*loc. cit. supra*, p. 464): "A skeleton, now in the Indian Museum, Calcutta, measures 11 feet 3 inches, so the animal when living, if the skeleton is correctly mounted, must have been nearly 12 feet high." But in a foot-note Mr. Blanford adds: "Since the above was written, I have been told by Mr. Sanderson that he compared the femur of the Calcutta skeleton with that of an elephant known to have been less than 10 feet high, and only found one-eighth inch difference in length."

For many years I have heard of "Jung Pershad," the late Sir Jung Bahadoor's favourite elephant that was used in catching wild elephants in the Nepal Terai, as being probably the largest known living Indian elephant, but I do not know "Jung Pershad's" actual dimensions.

A female elephant from Mysore, called "Zebi," which lived in England, in the Zoological Garden at Clifton, near Bristol, from 1868 till 1910, when she died at the age of about fifty years, grew to a great size. Her head is now mounted in the Bristol Museum, and on the label it is stated that she "was believed to be the second largest Indian elephant known, her height being approximately 10 feet and her weight about 5 tons." Unfortunately the label does not state whether she was measured at the shoulder when alive.

3.—Size of Elephants' Tusks.

Mr. Blanford (*loc. cit. supra*, p. 464) writes: "Tusks vary greatly, the longest recorded I believe (Sir V. Brooke's from Mysore) measured 8 feet and weighed 90 lbs., but a shorter tusk from Gorakhpur is said to have weighed 100 lbs. Both were from elephants with but one tusk perfect. Two pairs from the Gáro hills are said to have weighed 157 and 155 lbs. respectively ('Asian,' October 16, 1888, p. 35)."

In the Trivandrum Museum I saw a Travancore elephant's tusk said to weigh 77 lbs. (34·92 kilogrammes).

In the "Journal of the Bombay Natural History Society," Vol. XI, No. 2 (1897), page 335, I gave some account of the collection of tusks then existing in the Royal Siamese Museum at Bangkok; the measurements of the four finest pairs were as follows :—

Length.				Circumference.		Notes.
Ft.	Ins.	Ft.	Ins.	Inches.	Inches.	
7	$4\frac{1}{2}$	7	$4\frac{1}{2}$	$16\frac{3}{4}$	$16\frac{3}{8}$	A massive, even pair.
7	$8\frac{1}{2}$	7	8	$13\frac{1}{4}$	$13\frac{1}{2}$	Points of tusks much worn.
8	3	8	4	$14\frac{1}{4}$	$14\frac{1}{2}$	A slender, symmetrical pair.
9	0	9	$10\frac{5}{1}$	$15\frac{3}{8}$	$15\frac{1}{2}$	These belonged to an elephant which is said to have died in Bangkok about the year 1877.

4.—Age of Elephants.

Mr. Blanford (*loc. cit. supra*, p. 446) writes: "An elephant is full grown, but not fully mature, at twenty-five years of age, and individuals have been known to live over 100 years in captivity; in a wild state their existence probably extends to 150 years." *

I much regret not to be able at present to agree with this statement. So far I have been unable to find any absolutely convincing evidence of an elephant living to the age of 100 years, and in regard to the last sentence it appears to me that it is probable that an elephant would live longer in captivity, carefully tended and guarded against accidents, than it would in a wild state, exposed to the many chances

---

* Elephants are mature at an earlier age than is commonly supposed. A pair of Asiatic elephants that were in my charge for nearly seven years became sexually adult when, we believe, between eleven and fourteen years old, and the female gave birth to a calf when she was about fourteen to sixteen years old. As far as my present observations go, African elephants reach maturity even earlier.

which terminate the careers of wild things once old age begins to affect their bodily power.

In each place where I saw elephants in India I made enquiries as to their age. At Trivandrum the elephant said to be oldest was a male, "Julius Cæsar"; his ascertained minimum age was thirty, and his supposed approximate age fifty-five years. The big male, "Chandra Sekaren," had an ascertained minimum age of twenty-five, and a supposed approximate age of forty years. At Alwar I saw an elephant believed to be seventy years old, but this the headman told me was a very great age for an elephant, and the actual record of this animal is only known for the last twenty-four years, his possible forty-six previous years are only estimated. A big male at Alwar had an ascertained minimum of thirty-two years, and a supposed approximate age of fifty-five.

It is interesting to compare the above statements with somewhat similar ones collected thirty-one years ago. In 1882, at the request of my father (the late Sir William Flower, K.C.B., President of the Zoological Society of London), the Government of Madras went into the question of the age of the elephants then living in their employ; it was reported on April 17, 1882, that there were:—

One elephant said to be aged 98 years.
Two elephants   ,,   ,,   95   ,,
One elephant   ,,   ,,   91   ,,
   ,,   ,,   ,,   ,,   90   ,,
   ,,   ,,   ,,   ,,   89   ,,
   ,,   ,,   ,,   ,,   80   ,,
Eighteen elephants said to be aged between 70 and 80 years.*

But in answer to further enquiries as to how the original age of these animals was found out, it was reported, on July 5, 1882, that their ages were only *estimated!*

So these figures do not really prove that an elephant in India lives longer than say fifty years, as appears to be the case in European menageries.

The ages of elephants in Madras in 1882 were "roughly

---

* Actually none of these elephants appear to have been, at that date, in the possession of the Madras Government for more than forty-three years.

judged of by the overturning of the upper lap of the ear. When turned down about one inch the elephant is supposed to be about thirty years old; when between one and two inches, the age ranged from thirty-six to sixty years; and when above two inches the elephant is considered aged."

The five oldest Asiatic elephants * that I have known personally in Europe were:—

(i) "Anton," a male, who lived in the Zoological Garden at Hamburg, Germany, from July 24, 1871, to October 26, 1907, *i.e.* thirty-six years three months two days, and was thirty-eight to forty years old at his death.

(ii) "Bella," a female, who arrived in the Zoological Garden at Cologne, Germany, on April 26, 1872, and lived there thirty-eight years nine months, and was forty-three years old at her death.

(iii) "Lilli," a female, who lived for forty-seven years in the Zoological Garden of Dresden, Germany, before she died in 1911.

(iv) "Zebi," a female, who arrived in the Zoological Garden at Clifton, England, November 4, 1868, and lived there for over forty-one years two months, and was apparently forty-nine or fifty years old when she died.

(v) "Suffa Cully," a female, received in the London Zoological Garden in May, 1876, and still alive there.

5.—Elephants' Face-glands.

At Alwar the upper gland on the face of an elephant is called "Daam," and the lower gland "Khamūka."

6.—Colour of Elephants' Eyes.

I noticed at Alwar that the colour of the irides varies much in individual elephants: females may have pale yellow irides, though dark brown is more usual.

---

* African elephants, so far as my present knowledge goes, are shorter-lived than Asiatic ones.

VII.—NOTES ON CROCODILES.

1. Species of Crocodiles found in India.
2. Geographical Distribution of Indian Crocodiles.
3. Shape of Indian Crocodiles.
4. Size of Indian Crocodiles.
5. Food of the Gharial.
6. Stones in Crocodiles' Stomachs.
7. Crocodiles in Indian Zoological Gardens.
8. "Tame" Crocodiles in India.
9. The Crocodiles of Jaipur.
10. The Crocodiles of "Mugger Pir," Karachi.
11. The Crocodiles of Kankeria.
12. The Crocodiles of Sarkej.

1.—Species of Crocodiles found in India.

Many otherwise well-educated people in India have very vague ideas as to the species of crocodiles found wild in that country. It may be as well to mention that true alligators, the genus *Alligator* of zoologists, only occur in parts of the United States of North America and in China, and that true crocodiles are also found in the southern parts of both those countries.

Two *genera* of Crocodilians exist in India, which Mr. G. A. Boulenger, F.R.S., thus distinguishes :—

Garialis. Snout extremely narrow and elongate ; twenty-seven teeth or more on each side of upper jaw.
Crocodilus. Snout moderate ; sixteen to nineteen teeth on each side of upper jaw.

The snout being the portion of the head in front of the orbits.

Of the first genus there is only a single species :—
The Gharial, *Garialis gangeticus.*

Of the second genus there are two species :—
The Estuarine Crocodile, *Crocodilus porosus.*
The Marsh Crocodile, or Mugger, *Crocodilus palustris.*

Mr. Boulenger ("Fauna of British India," Reptilia, 1890, pp. 4, 5) points out the following differences between these two species :—

C. porosus.

(1) Seventeen to nineteen upper teeth on each side.
(2) Snout $1\frac{2}{3}$ to $2\frac{1}{4}$ times as long as broad at the base.
(3) Snout with a more or less strong ridge on each side in front of the eye.
(4) Four teeth in each praemaxillary bone (in adults).
(5) Praemaxillo-maxillary suture, on the palate, directed backwards or W-shaped.

(6) Postoccipital scutes usually absent, sometimes small and irregular.
(7) Dorsal scutes forming four to eight longitudinal series.

(8) Dorsal scutes, in a transverse series, separated from each other by the leathery skin.

C. palustris.

(1) Nineteen upper teeth on each side.
(2) Snout $1\frac{1}{3}$ to $1\frac{1}{2}$ times as long as broad at the base.
(3) Snout without any ridges.

(4) Five teeth in each praemaxillary bone.
(5) Praemaxillo-maxillary suture, on the palate, transverse, nearly straight, or curved forwards.

(6) Postoccipital scutes small, consisting of two pairs in a transverse series.
(7) Dorsal scutes forming four, rarely six, longitudinal series.

(8) Dorsal scutes, in a transverse series, suturally united to each other.

(*Vide* Boulenger, "Fauna of the Malay Peninsula," Reptiles, 1912, p. 5.)

2.—Geographical Distribution of Indian Crocodiles.

Garialis gangeticus.

" Indus, Ganges, and Brahmaputra rivers and their larger tributaries ; also Mahánadi of Orissa, and Koladyne river, Arrakan, but not the Nerbudda, Tapti, Godávari, Kistna, Irrawaddy, or other rivers of India or Burma." (Boulenger, "Fauna of British India," Reptilia, 1890, p. 3.)

Crocodilus porosus.

Widely distributed in the East Indies ; Malabar Coast, Ceylon, East coast of India, Bengal, Burma, Siam, South China, Malay Peninsula and Archipelago, Philippines, New Guinea, North Australia, Solomon Islands, Fiji Islands.

" Entering salt water and frequently occurring out at sea. It is not certain that this species is found far above the tideway in rivers." (Boulenger, *loc. cit. supra*, p. 4.)

Crocodilus palustris.

" India, Ceylon, Burma, Malay Peninsula and Archipelago. This is the common Crocodile of India, found in rivers, marshes, and ponds, and extending west to Sind and Baluchistan." (Boulenger, *loc. cit. supra*, p. 5.)

No absolutely unimpeachable evidence appears to exist of its occurence in the Malay Peninsula or Archipelago.

As to the occurrence of *C. palustris* in Baluchistan, Prof. Dr. Erich Zugmayer, of the Munich Museum, kindly told me that he himself had obtained specimens near Las Bela, and near Bassali, and even as far west as the Dascht River, which is near the frontier of Persia.

3.—Shape of Indian Crocodiles.

Garialis gangeticus.

The gharial has an extremely narrow and elongate snout, slightly dilated at the end.

" A strong crest on the outer edge of the forearm, leg, and foot." (Boulenger, *loc. cit. supra*, p. 3.)

Crocodilus porosus.

The estuarine crocodile resembles the African *Crocodilus niloticus* in shape of snout and general build.

" A serrated fringe on the outer edge of the leg." (Boulenger, *loc. cit. supra*, p. 4.)

Crocodilus palustris.

Adult muggers have a short broad snout, suggesting the shape of that of the American *Alligator mississippiensis.*

" A serrated fringe on the outer edge of the leg." (Boulenger, *loc. cit. supra*, p. 5.)

Till it attains a length of about six feet (1·82 metres) *C. palustris* has the same general slender build as *C. niloticus* or *C. porosus*, but after that it tends to become a very heavy thick-set animal like the American *Alligator*.

4.—Size of Indian Crocodiles.

THE GHARIAL, *Garialis gangeticus*.

" The gharial reaches a length of 20 feet." (Boulenger, "Fauna of British India," Reptilia, 1890, p. 3.)

In the Victoria Museum, Karachi, I saw a stuffed gharial from the Indus, in length, as stuffed, 14 feet 9 inches.[*]

In the Natural History Museum of Vienna there are a pair of gharials stuffed ; the male about 5·50 metres (18 feet ½ inch), the female about 5 metres (16 feet 5 inches) in total length.

Captain J. Johansen has told me of gharials reaching the length of 25 to 30 feet (7·62 to 9·14 metres) in the Brahmaputra River. These are referred to in the late Herr Carl Hagenbeck's book, " Beasts and Men," 1909, on page 201, and I have myself seen very large individuals in the Ganges. The largest gharial that I ever saw was in that river, near Ghazipur, in January, 1895. He successfully eluded my endeavours to secure him as a specimen ; he was a male, and according to the estimates, made very carefully at the time, was *not less* than 30 feet (9·14 metres) in length, and may have been several feet longer.[†]

THE ESTUARINE CROCODILE, *Crocodilus porosus*.

In the British Museum there is a skull from Bawisal, Bengal, "stated by the donor to have pertained to a specimen 33 feet long, and measuring 13 feet 8 inches round the body." (Boulenger, "Brit. Mus. Cat." Chel., 1889, p. 285.)

A specimen killed at Matang, Perak, Malay Peninsula, measured in length 24 feet 8 inches (7·51 metres). (*Vide* P.Z.S., 1899, p. 624.)

---

[*] Mr. F. Ludlow kindly measured this specimen for me, and said that, allowing for shrinkage, the length would have been " undoubtedly greater in the freshly killed animal."

[†] Lieut. T. G. Carless recorded having seen in 1836–37, in the River Indus, four gharials *at least 30 feet long*. *Vide* " The Zoological Gardens, Karachi," by H. P. Farrel and F. Ludlow, 1913, page 36.

In the Indian Museum, Calcutta, there are two large skulls of this species, one with its skeleton.*

In the Madras Museum there is a skeleton of a *Crocodilus porosus* "10½ feet (3·20 metres) long. Fort St. George Moat. 1889." Dr. J. R. Henderson has been so kind as to give me the following dimensions of the skull of this specimen :—

Length, from a point in line with the top of the snout to one in line with the posterior border of the occipital condyle, 1 foot 6½ inches (0·469 metre).
Length, including lower jaw, 1 foot 10½ inches (0·571 metre).

Personally I have never met a really big *Crocodilus porosus* alive : the largest specimen I have seen in the flesh was about 12 feet (3·67 metres) long ; this was in the Kedah River in the Malay Peninsula.

The six largest skulls of this species that I have notes of as having actually measured are :—

(i) A skull from the Tacheen River, Siam (in the Royal Siamese Museum, Bangkok) :—

Length, without lower jaw, about 2 feet 11½ inches (*ca.* 0·901 metre).
Breadth in front of orbits, following curve, about 1 foot 5½ inches (*ca.* 0·444 metre).

(ii) A skull, locality unknown (in the Raffles Museum, Singapore) :—

Length (with lower jaw ?), about 3 feet 2½ inches (*ca.* 0·977 metre).
Breadth in front of orbits, following curve, about 1 foot 5¼ inches (*ca.* 0·438 metre).

---

* Mr S. W. Kemp has been so kind as to send the following measurements of these specimens, and to inform me that the localities from whence they came are unknown.

	Ft.	Ins.	Ft.	Ins.
Total length of skeleton	14	11	—	—
Skull, total length, without lower jaw	2	3	2	6
„ „ „ with „ „	2	9	3	2
„ breadth in front of orbits, following curves	1	4½	1	5½
„ „ „ „ „ in straight line	1	½	1	3½

(iii) A skull from Pahang, Malay Peninsula (in the Taiping Museum, presented by Mr. G. F. W. Curtis) :—

Length, including lower jaw, about 2 feet 11½ inches (*ca.* 0·901 metre).

Breadth in front of orbits, following curve, about 1 foot 3¼ inches (*ca.* 0·387 metre).

(iv) A skull from Borneo (in the Manchester Museum, presented by Mr. R. D. Darbishire) :—

Length, without lower jaw, about 2 feet 5½ inches (*ca.* 0·749 metre).

Length, with lower jaw, about 2 feet 10 inches (*ca.* 0·864 metre).

(v) A skull from a river in the interior of Borneo, collected by M. S. Müller in 1831 (in the Leyden Museum) :—

Length, without lower jaw, about 2 feet 4¾ inches (*ca.* 0·730 metre).

Length, with lower jaw, about 2 feet 9 inches (*ca.* 0·838 metre).

Breadth in front of orbits, following curve, about 1 foot 4¼ inches (*ca.* 0·412 metre).

(vi) A skull, without locality but probably from Java (in the Munich Museum) :—

Length, without lower jaw, about 2 feet 3⅕ inches (*ca.* 0·691 metre).

Length, with lower jaw, about 2 feet 6¾ inches (*ca.* 0·781 metre).

Length in median line of skull, about 2 feet 0⅗ inches (*ca.* 0·620 metre).

Breadth in front of orbits, following curve, about 1 foot 1⅓ inches (*ca.* 0·339 metre).

In the Trivandrum Museum I saw a coloured plaster cast of an Estuarine Crocodile, *Crocodilus porosus*, 14 feet 4 inches (4·36 metres) long. "This Crocodile was harpooned by one Mathai Chakko and some Valens in the Cheppanum Canal to the south-east of Tripunathura in the Cochin State,

and was after an exciting chase of three hours caught alive
at 10 a.m. on the 1st August, 1904. Though three of
the harpoons discharged hit it, one in the eye, it lived for 30
hours after it was landed. For about two years before it
was caught, it was the terror of the Cheppanum Canal having
killed several heads of cattle, goats and dogs. Though it
attacked and mauled several men, only one or two of them
actually died." (Report, Trivandrum Museum, 1904–1909,
p. 6.)

THE MARSH CROCODILE, OR MUGGER, *Crocodilus palustris.*

"Grows to a length of 15 feet, or more." (Boulenger,
"Fauna of the Malay Peninsula," Reptiles, 1912, p. 6.)

In the Victoria Museum at Karachi there is a stuffed
Crocodilus palustris from Jungshai, Sind, presented by
Mr. F. Gleadow. · This animal was in life probably about
13 feet (3·96 metres) long.

In the Madras Museum there is a large skull of this species,
which measures :—

Length, without lower jaw, 1 foot 8 ¾ inches (0·527 metre).
Length, with lower jaw, 2 feet 1 inch (0·635 metre).

Dr. J. R. Henderson very kindly had this specimen
measured at my request ; he writes : " The length of the
crocodile skull (first measurement) is in the middle line
from a point in· line with the tip of the snout to one in line
with the posterior border of the occipital condyle."

The largest specimens of *Crocodilus palustris* that I have
seen alive were in the Ganges, downstream of Benares, in
January, 1895 ; some of these attained to an extraordinary
girth of body and stoutness of limbs and tail, but as I never
succeeded in catching or killing one of these monsters I am
not in a position to give the actual dimensions to which
they attain.

5.—Food of the Gharial.

It is frequently believed in India, and has been stated in
books, that the Gharial, *Garialis gangeticus*, feeds entirely
upon fish. This is, however, not the case. I have heard
of about three undoubted instances of men being killed by

gharials, but have only at present the data of one case. In 1897 a gharial killed and ate a fakir who was bathing in the river at Mahaban, near Muttra; about a week later, on April 19, 1897, the reptile was shot by Captain E. O. Wathen, 5th Royal Irish Lancers, and some remains of the fakir found inside it. The gharial measured 15 feet 10 inches (4·82 metres) overall in a straight line, and was in girth, one foot behind the forelegs, 5 feet 8 inches (1·72 metres). The skeleton of this gharial and a bone of the man it killed are now in the British Museum (Natural History).

I am indebted to Mr. G. A. Boulenger, F.R.S., for his kindness in looking up the British Museum registers to confirm the above details, and for giving me other valuable and interesting information about the Crocodilia.

6.—Stones in Crocodiles' Stomachs.

In the Indian Museum at Calcutta some stones found in the stomach of a gharial are exhibited.

Two recent papers on this subject may by mentioned :—

(1) H. W. Forsyth, " Stones in Gharials' Stomachs." (Journal Bombay Nat. His. Society, XX, 1910, p. 229).

(2) E. W. Shann, " The Contents of the Stomach of a Nile Crocodile." (Cairo Scientific Journal, IV, 1910, p. 279.)

In " Land and Water," August 24, 1896 (or 1898 ?), there is an account of the shooting on the Gunduk river, in Tirhoot, of a gharial 16 feet 11 inches (5·15 metres) in length, apparently a male, which states : " his inside was full of stones, varying from 3 inches in diameter to large gravel, a very useful digestive, but heavy."

7.—Crocodiles in Indian Zoological Gardens.

In the twelve zoological gardens visited I only saw forty-two Crocodilians, twenty-one of these being in Calcutta, where four species were represented, *Garialis gangeticus* (five individuals), *Crocodilus porosus*, *Crocodilus palustris*, and the American *Alligator mississippiensis*.

The remaining twenty-one were all true crocodiles and were thus distributed :—

BARODA.   Two, one large and one of medium size.   Owing to the construction of the cage it was difficult to see these animals clearly ; they appeared to be specimens of *Crocodilus palustris*, with unusually long snouts.

BOMBAY.   One small *Crocodilus palustris.*

KARACHI.   Three *Crocodilus palustris.*   One of these is said to have been fourteen years there ; it is now about 6 feet (1·82 metres) long, It is, and always has been, very fierce. It once escaped from its enclosure into the waterfowl pond, where it seized and killed a black swan. The two other specimens (which are now at Giza) were caught near Ghimpir, about seventy-four miles from Karachi. The Sindhi name for this species is " Wagu."

MADRAS.   Five.   The largest, about 6 feet (1·82 metres) long, appeared to be a *Crocodilus palustris.*   A *Crocodilus porosus* was about 5 feet (1·52 metres) in length.

MYSORE.   Five.   All from the Cauvery River, in Mysore territory.   These are believed to be *Crocodilus palustris.*

RANGOON.   One small *Crocodilus porosus.*

TRIVANDRUM.   Four *Crocodilus palustris,* three of which are of special interest from the length of time they have lived in the Trivandrum Garden.   The two largest are estimated to be about 10 feet (3·04 metres) in length, and are nice and fat; the third specimen is between 5 and 6 feet (1·52 to 1·82 metres) long and of slender build.   The dates these animals were obtained, according to the official register, are :—

May 11, 1882, *i.e.* thirty years eleven months fifteen days, and still alive (April 26, 1913).

December 24, 1885, *i.e.* twenty-seven years four months two days and still alive (April 26, 1913).

November 26, 1892, *i.e.* twenty years five months and still alive (April 26, 1913).

8.—" Tame" Crocodiles in India.

The late Mr. John Lockwood Kipling, C.I.E., in his book, " Beast and Man in India" (1904), on page 318, writes :—
" Crocodiles are occasionally regarded as sacred, one cannot say kept and periodically fed. *Muggur pir* near Karachi is a pond full of these creatures, which are often fed for the amusement of visitors. There is a legend of a British officer who crossed this pool, using its inhabitants as stepping-stones in his daring passage.* In some of the lakes in Rajputana they are cherished and come to the Brahman's call ; not one may be visible at first, but there is first a ripple, then a slow, hideous head protrudes, then another, till the water is alive with crocodiles."

I made very many enquiries as to the places in India in which crocodiles are thus kept and will come to a man's call, but could only hear of the Jaipur tank and the Karachi Mugger Pir, at both of which places the crocodiles can hardly be considered sacred, their food being mainly if not entirely provided by the money of European visitors, and it is noteworthy that in both places the keepers of the reptiles are Mohammedans and not Hindoos.

Mr. N. B. Kinnear, of the Bombay Natural History Society, who had been particularly kind in trying to obtain information about crocodiles for me, called my attention to an article in " The Times of India," of June 7, 1913 (the day I left India), entitled " Crocodile Habits," quoted from a paper in the " Cornhill Magazine" by Mr. Shelland Bradley. Mr. Bradley describes having seen at a tank known as the Ghoradighi, " constructed by the great Khan Jahan Ali four hundred and fifty years ago," in the Sunderbans, crocodiles which when called by one of the villagers come right up into the shallow water below the bank and wait to be fed with fowls. These crocodiles are " said to be the descendants of those placed there by Khan Jahan Ali," and so would be of Mohammedan interest as curiosities and have nothing to do with Hindoo worship.

---

* I believe this feat was accomplished in, or about, the year 1842. The officer was Lieut. Beresford of the 86th Foot, now the 2nd Battalion Royal Irish Rifles. *See* Burton, " Sind," 2nd Edition, 1851, Vol. I, page 56, and also Burton, " Sind Revisited," 1877, Vol. I, page 99.

9.—The Crocodiles of Jaipur.

Beyond the garden of the Palace in Jaipur there is a square tank where crocodiles are kept. I visited this place on May 17, 1913. I could not measure the area of the tank, but it cannot be less than five acres (2·02 hectares) and might perhaps be as much as twelve acres (4·85 hectares) in extent. This tank is said to have been built about 200 years ago, the city of Jaipur having been founded by Jai Singh in 1728.

So far as I could ascertain, there are now nine crocodiles in the tank, apparently all *Crocodilus palustris*. The largest specimens appear to be about 9 to 10 feet (2·74 to 3·04 metres) in length and are very stout. Apparently they are very old and have scarred and lumpy heads. One individual is toothless.

The keeper of the crocodiles, a Mohammedan, took me to a flight of stone steps leading down into the water and then commenced calling. It was a wonderful sight to see the crocodiles, hundreds of yards away in various parts of the large tank, take notice of the man's voice and start swimming to the steps ; and not only crocodiles came but also several huge soft-turtles, with longitudinal streaks on their heads, probably *Trionyx gangeticus*.

Actually five crocodiles swam to the steps and were very bold, coming out of the water to half or three quarters of their length. In fact, probably they would have come right out of the tank and up the steps if their keeper had allowed them to do so.

The man fed them by hand with pieces of raw meat. The crocodiles show much individuality : one just holds its mouth wide open and keeps still, waiting its turn to be fed, another fusses about snapping and hissing and roaring.

The keeper knows each reptile apart and its character. He carries a thin stick in his hand with which he smacks his charges over the head and makes them behave and take their portions of the meal in due order.

Meanwhile the turtles do all they can to attract notice, and the keeper with his stick and voice drives off the crocodiles and gives the turtles their share.

As these turtles snap so furiously, their pieces of meat are fastened to a string hung from the end of a long stick and so handed out to them.

10.—The Crocodiles of "Mugger Pir," Karachi.

Pir Mangho, the tomb of Haji Mangho, is commonly called "Mugger. Pir" (pronounced "Mugger Peer") on account of the crocodiles living in the tank at the foot of the tomb (*see* Pl. III). It is about ten miles from the Karachi Post Office, and, I am told, about six miles east of the Hub River, the frontier between Sind and Baluchistan.

In "The Karachi Handbook" (Sind Gazette Press), 1913, on page 58, we read : "Pir Mangho has undergone that curious metamorphosis which is exemplified in the names of so many English public houses and become Mugger Peer, which will be its name when the last of the "muggers" is a specimen in the Karachi Museum. It is the tomb of HAJI MANGHO, a holy hermit who is said to have settled there about the middle of the thirteenth century."

Haji Mangho was an Arab (*vide* "Karachi Guide and Directory," 1913, p. 85). His tomb is a place of pilgrimage for Mohammedans. "It is also a resort of Hindu devotees, who call it LALA JASRAJ" (*vide* "Karachi Handbook," *loc. cit. supra.*).

From the "Karachi Handbook" I also learn that Lieut. Carless of the Indian Navy visited the place in 1838 and wrote :—

"The swamp is not more than 150 yards long by about 80 yards broad, and in this confined space I counted above 200 large ones (crocodiles), from 8 to 15 feet long, while those of a smaller size were innumerable."

Capt. E. B. Eastwick visited Karachi about 1841 and went to "the Magar Talāo," as he calls it, in August or September. In his book, " Dry Leaves from Young Egypt," 3rd Edition, 1851, page 218, he writes that he saw : "at least three score huge alligators, some of them fifteen feet in length." In the copy of this book in the Karachi General Library in the Frere Hall (which the Librarian very

kindly allowed me to see) some one has noted " 200 " for
" three score " and " 12 " for " fifteen."

Sir Richard F. Burton, then a Lieutenant in the Bombay
Army, must have been here about 1842. In his book,
" Scinde, or the Unhappy Valley," 2nd Edition, 1851, Vol. I,
page 50 (for seeing this rare book I am indebted to the
officials of the Royal Geographical Society, London), he
writes :—

" The little bog before us, though not more than a hundred
yards* down the centre, by half that breadth, contains hundreds of alligators of every size from two to twenty feet."

Mr. B. B. Woodward, of the British Museum (Natural
History), has been so good as to call my attention to
Burton's later book, " Sind Revisited," which was published
in 1877, and in which from pages 88 to 103 there are
numerous references to Mugger Pir. And the legend of its
origin is given, how Haji Mangho settled in this barren
spot and prayed for thirteen years and then caused a rill to
trickle from the rock, and how his four disciples also wrought
miracles and thus made the oasis. Lal Shahbaz made the
hot spring, Jimal el Din turned his tooth-brushing stick
into a date palm, Jelal Jaymaya made the tree rain down
honey and butter, and, what most concerns us, Farid el Din
turned a flower into a crocodile!

But, turning from legend to history, we find that between
the dates of Burton's two visits the methods of keeping the
crocodiles had changed. Formerly, as we know from
Carless, they lived at liberty in a natural swamp, but as
they used to stray, attack people, or steal children, and
probably also owing to the increased value of land in the
oasis, the swamp was drained and the crocodiles confined to
a tank surrounded by a high wall. This evidently led to a
great reduction in the number of the reptiles, as instead of
the " hundreds " of his first visit, Burton puts the number
at forty on his second visit, these being all big ones. He
also mentions that " Mor Sahib," the big crocodile (said
to have been eighteen feet long) referred to in several books,
had died before his second visit.

* In " Sind Revisited," 1877, Vol. I, p. 97, Burton writes " 400 feet."

I visited Mugger Pir on May 27, 1913, and found El Sheikh Mutka Glamali in charge of the place. He is probably a son or nephew of the Mujáwir, Miyan Mutka, that Burton mentions.

The oasis where Haji Mangho lived is surrounded by desert hills and stoney valleys with a scattered bush vegetation. The desert sand appears to be encroaching on the cultivated portion of this oasis.

The enclosure in which the crocodiles are now kept is an irregular pentagon about seventy paces in circumference; in it is a tank of water and some trees. It is surrounded by a high wall.

Inside this enclosure I saw twenty-four adult Marsh Crocodiles, or Muggers, *Crocodilus palustris*, whose length I estimated to be from about 6 feet (1·82 metres) to about 9 feet (2·74 metres), and also three young crocodiles, of the same species, each about 18 inches (0·457 metre) in length. Some of the adults had very battered jaws and irregular teeth.

The sheikh told me that there are seventy to eighty young crocodiles in the tank: this may or may not be so.

The adults apparently lay their eggs in a sandy corner of the enclosure.

Two goats were killed to provide the crocodiles with a meal, and Sheikh Mutka Glamali and I got over the wall into the enclosure and stood at the edge of the pool while he called the reptiles. A few took no notice at all of his voice, but the majority hurried to us, crawling over each other in the shallow water in their eagerness to catch the pieces of meat thrown to them. There appeared to be no method in the feeding. It was no orderly meal but a wild scramble, crocodiles snapping, fighting, and rolling about with much splashing of dirty water.

An incident which helped one to realize the great strength of a crocodile's jaws was an animal, not more than seven or eight feet in length, seizing the severed head of a goat and while holding it high above the water crushing it to pieces, with a horrid sound of breaking bones, as easily as a man might break a hen's egg between his hands.

11.—The Crocodiles of Kankeria.

In the Kankeria Tank near Ahmedabad (*see* p. 34) on
May 30, 1913, I saw two crocodiles swimming about; they
were large enough to be dangerous, but in spite of that
many people were bathing, though at the further end of the
tank from where I saw the crocodiles. The gardener on the
island Nagina told me that there were not more than four
crocodiles in the tank, but that they were all big ones.
These crocodiles are probably *Crocodilus palustris·*

12.—The Crocodiles of Sarkej.

Murray ("Handbook to India," 8th Edition, 1911, p. 130)
in describing Sarkej (Sarkhej) writes : " Numbers of people
bathe in the tank in spite of the alligators."
On May 31, 1913, I visited Sarkej, near Ahmedabad
(*see* p. 35). The people who live near there assured me
that there are now no crocodiles in the tank. I think this
is true. I spent over an hour at the tank, the water was
very shallow, and I saw a herd of buffaloes driven into the
water, where they wallowed, and there was never a sign of
a crocodile.

The people told me that formerly there were crocodiles
here, but as the area of the water in the tank decreased and
the water got shallower and shallower the crocodiles disap-
peared (probably by migrating to other waters) ; all except
a few big crocodiles that remained at Sarkej, and these were
shot by the Sahib-lôg.

VIII.—NOTES AT AHMEDABAD.

The Pinjrapol.

The Pinjrapol, or Asylum for Animals, is an institution of a religious character maintained by the Jains in those Indian towns where this sect is numerous and wealthy. Hearing that the Pinjrapol of Ahmedabad was the principal of these asylums, I visited it on May 30, 1913. It consists of several courtyards and sheds in which an enormous number of animals are herded together, apparently in a very promiscuous manner. If attempts are made to keep the place clean, they are not noticeable. It was a very, very sad sight to see the large number of aged or crippled animals which are here kept alive to the last, instead of being " put out of their misery," as would be said in England. But there is more to be learnt here than is apparent at a first glance. Theoretically it is to European ideas a place of sad cruelty to animals, not intentional cruelty, but misdirected kindness on the part of the pious supporters. With my preconceived notions of how such animals (I will not enter into details of their wounds or other ills) should behave, it was with no little fear of being gored or kicked that I ventured among them, but they were quiet, friendly with the old man who tended them, and appeared to have absolute confidence in the presence of human beings. The majority of the mammals were domestic oxen and buffaloes ; there were also sheep, goats, horses, and dogs. A white rat was running about loose and tame enough to allow itself to be picked up in the hand. In one place there were a few Monkeys, *Macacus rhesus* and *Macacus cynomolgus* ; in another a Hedgehog, *Erinaceus sp.* (an animal that I did not see in any of the Indian Zoological Gardens). A Gazelle, *Gazella benneti*, was wandering at large, and two male Nylgai, *Boselaphus tragocamelus*, were tethered in a shed.

Large numbers of domestic or semi-domestic birds, such as pigeons, peafowl, poultry, and ducks, inhabit the Pinjrapol and have a free happy existence, but the same can hardly be said of the birds, principally parrakeets and birds of prey, which are kept in cages.

Murray ("Handbook to India," 8th Edition, 1911, p. 127), in mentioning the Ahmedabad Pinjrapol places the number of animals lodged in it as about 800, and adds : " There is also a room where insects are fed." I enquired about this at Ahmedabad, and took the advice locally given to me that I was on no account to go to this room: I was *told* that the method of feeding the insects is to pay a man a certain sum to allow the parasites to feed on his blood for a fixed number of hours. The man is then strapped down on a bed and left in the room and not allowed out till the time bargained for has expired.

Kankeria Tank.

The "Houz-i-Kootub," Lake of Pebbles, or Kankeria Tank, near Ahmedabad in Guzerat, is said to be about 72 acres (29·13 hectares) in area (*see* Murray, "Handbook to India," 8th Edition, 1911, p. 128). Kootboobeen Shah, King of Ahmedabad, commenced to build it in 1452 A.D. (856 A.H.), and Mr. A. A. Borradaile, Collector, restored it in 1879 A.D.

In the tank is the island Nagina, the Gem, connected to the mainland by a wooded causeway and a beautiful stone bridge. The island is laid out as a garden, rich with sub-tropical and tropical vegetation, Casuarina, Dôm Palm, Cocoanut Palm, Mango, Arbor vitae, Jasmine, Bougainvillea, Neem, Madagascar Papyrus, etc., etc., and is a sanctuary for birds.

This island is indeed a wonderful place for seeing birds, both for the numbers that frequent it and for their tameness and want of fear of mankind. The most marvellous sight is just after sunset when thousands of birds fly in from all directions to roost on the island, and gradually settle down for the night. It is not only a wonderful thing to see but also to hear. There are Jungle Crows, *Corvus macrorhynchus*, House Crows, *Corvus splendens*, Black-headed Bulbuls, *Molpastes haemorrhous,* great crowds of practically tame Mynas, *Acridotheres tristis*, many Kingfishers both of the large pied species, *Ceryle varia,* and of the Indian form of the common European species, *Alcedo ispida,* Koels, *Eudynamis honorata,* Crow-Pheasants, *Centropus sinensis,* flocks of Ring-necked Parrakeets, *Palaeornis torquatus,* White-breasted Waterhens, *Amaurornis phoenicurus,* many very confiding Red-wattled Lapwings, *Sarcogrammus indicus,* Purple Herons, *Ardea manillensis,* Grey Herons, *Ardea cinerea,* Pond Herons, *Ardeola grayi,* Little Green Herons, *Butorides javanica,* Night Herons, *Nycticorax griseus,* and Chestnut Bitterns, *Ardetta cinnamomea.* Perhaps the most striking sight of all is to see flock after flock of either black or white birds fly in, the black flocks being composed of Large Cormorants, *Phalacrocorax carbo,* Little Cormorants, *Phalacrocorax javanicus,* Snake-birds, *Plotus melanogaster,* and Glossy Ibis, *Plegadis falcinellus,* while the white flocks comprise the Large Egret, *Herodias alba,* Smaller or Little Egrets, *Herodias intermedia* or *Herodias garzetta,* and Eastern Cattle-Egrets, *Bubulcus coromandus.*

It was on May 30, 1913, that I visited the Kankeria Tank. I saw two crocodiles and several Soft-Turtles, *Trionyx.* Some of these turtles were very large, and one at least was of gigantic size. I saw it lying out on the bank of the causeway that leads to the island. The turtle was so big and so fat that I had difficulty in believing it to be real. I saw it slowly stretch out its head and then crawl deliberately into the water and leisurely swim away. It was impossible to estimate its size or weight, I can only say that I had no idea that a *Trionyx* ever grew so big or so *thick through,* although I have over eighteen years' experience of turtles of this genus in the Ganges, the Malay Peninsula, Siam, and the Nile.

Fishing is prohibited in the Kankeria Tank. The water appears to swarm with fish, which doubtless provide abundant food for the water-birds, crocodiles, and turtles.

Sarkej.

Sarkej, or Sarkhej, six or seven miles from Ahmedabad, in Guzerat, was visited on May 31, 1913.

King Mahmud Bigara, who reigned in Guzerat from 1459
to 1513 A.D., built the tank here, which is said to be
17½ acres (7·08 hectares) in area (*see* Murray, "Handbook
to India," 8th Edition, 1911, p. 130).

The water is now reduced to one large shallow pool in
the centre of the great rectangular place, the sides of which
are formed by tiers of stone steps, leading up to ruined
palaces and kiosks of great architectural beauty, which are
nowadays inhabited by large numbers of big grey Langur
Monkeys, *Semnopithecus entellus*, which run before one from
room to room, scamper over the roofs, and watch one
through windows and doorways.

Besides these monkeys I saw many little striped palm-
squirrels, a hare, countless birds, and some fine lizards
(of the genus *Calotes*) among the ruins.

The most noticeable of these wild birds were the Indian
House-Crow, *Corvus splendens*, the Babbler, *Argya caudata*,
the Black-headed Bulbul, *Molpastes haemorrhous*, the White-
eared Bulbul, *Molpastes leucotis*, the King-Crow, *Dicrurus
ater*, the Black-headed Myna, *Temenuchus pagodarum*, the
Myna, *Acridotheres tristis*, the Black Indian Robin, *Thamnobia
cambaiensis*, the Indian Silver-bill, *Uroloncha malabarica*,
the Indian House-Sparrow, *Passer domesticus indicus*, large
numbers of the very pretty Sykes's Striated Swallow,
Hirundo erythropygia, a black and white Wagtail, *Motacilla sp.*, the Ashy-crowned Finch-Lark, *Pyrrhulauda grisea*,
the Indian Roller or "Blue Jay," *Coracias indica*, the Green
Bea-eater, *Merops viridis*, a small brown Swift, apparently
Tachornis batassiensis, the Koel, *Eudynamis honorata*, the
Crow-Pheasant, *Centropus sinensis*, the Ring-necked Parra-
keet, *Palaeornis torquatus*, the Indian Pariah Kite, *Milvus
govinda*, large numbers of Blue-Rock Pigeons, *Columba livia*
or *Columba intermedia*, many Indian Palm-Doves, *Turtur
cambayensis*, and Laughing-Doves, *Turtur risorius*. Brilliant
Peacocks, *Pavo cristatus*, and stately Sarus Cranes, *Grus
antigone*, walked where they willed in these gardens of dead
kings. The cheery "Did-you-do-it" Plovers, or Red-
wattled Lapwings, *Sarcogrammus indicus*, ran here and
there on the banks of the tank, where the Pond-Herons,
Ardeola grayi, stood motionless, and White-necked Storks,

Dissura episcopus, and Black-headed White Ibises, *Ibis mela-nocephala,* came and went. Finally, and finest sight of all, must be mentioned a flock of thirty-eight individuals of the Painted Stork, *Pseudotantalus leucocephalus.*

The crocodiles of Sarkej, which were the object of my visit to the tank, I found to be extinct. However, the monkeys and the wild birds still make Sarkej a most attractive and interesting place for a zoologist to visit.

IX.—NOTES AT ALWAR.

Although there is no zoological garden at Alwar, there is
a very beautiful public garden, called the "Company Bagh,"
in which there is a stone house containing three live tigers,
a large cage with three black Himalayan bears, and, in the
fine greenhouse, tanks with countless gold-fish. Further-
more, in H.H. the Maharaja's private garden, the "Mangel
Bagh," near the Lansdown Palace, is the very remarkable
building known as the "Tiger House" (which is described
below) at present occupied by a fine pair of leopards.

Quantities of pretty little striped Palm-Squirrels, *Sciurus
palmarum*, play about the gardens of Alwar, and in the
country outside the city are very large numbers of Blackbuck,
Antilope cervicapra : I saw a female antelope of this species
wandering about the bazaars. In the evening many Flying
Foxes, *Pteropus medius*, are to be seen.

All birds are protected in Alwar, and the results of this
protection are beautiful. Not only are there very large numbers of birds to be seen, but they are so easily seen, as they
have confidence in mankind instead of fear; in fact the wild
birds are almost tame. I was at Alwar from May 17
to 19, 1913, and noted twenty-eight species of birds as
being "common" there, besides Crows, Babblers, Bulbuls,
King-Crows, Woodpeckers, Rollers, Bee-eaters, Kingfishers,
Hoopoes, Parrakeets, Kites, Doves, and Pond-Herons, I
would specially mention the Myna-birds, *Acridotheres tristis*,
the Jungle Mynas, *Aethiopsar fuscus*, the Pied Mynas,
Sturnopastor contra, the black Indian Robins, *Thamnobia
cambaiensis*, the Magpie-Robins, *Copsychus saularis*, the
Weaver birds, apparently *Ploceus bengalensis*, the great
flocks of Blue Rock Pigeons, *Columba livia* or *Columba
intermedia*, the Red-wattled Lapwings, *Sarcogrammus indicus*,
and especially the Peafowl, *Pavo cristatus*, the cocks with
lovely trains, which were everywhere, on the roads and on the
roofs, literally in thousands.

The Alwar Tiger House.

This enclosure was built by the late Maharaja of Alwar; it is so arranged that when driving to the palace or walking in the park one can see the wild animals, apparently at liberty. The animals actually live on a flat plateau, circular in plan, 75½ feet (23·03 metres) in diameter. This plateau is furnished with a sunk retiring room, a small pond of water, and a tree, a Banyan, *Ficus bengalensis* (apparently). The tree is pollarded with due discretion, so that, while giving shade to the animals, no broken branch may form a bridge by which they could escape.

The plateau is surrounded by a ditch, 18 feet (5·48 metres) in depth, and 30 feet (9·14 metres) in width.

Access for the keepers to the plateau is arranged for by a door in the garden opening on to a staircase leading underground to a passage which passes below the ditch and ascends by another staircase to the centre of the plateau, with of course the necessary iron gates to exclude the animals. Both the platform and the underground passage and staircase are lighted by electricity. At the time of my visit the underground passage was full of water and so could not be used.

A hedge, planted in a circle round the outside of the ditch 22 feet (6·70 metres) from its outer edge, serves to hide the ditch from the view of visitors walking in the garden and completes the illusion of the tigers or leopards being at large.

The walls of the ditch are finished in smooth-faced cement. The sunk retiring room in the centre of the plateau is surmounted by a small open-sided cage, the roof of which forms a bench for the animals, and above it is a pole carrying two powerful electric lamps.

In case the animals fall into the ditch, an opening has been constructed in the lower part of the outside wall of the ditch leading into a chamber, which can be closed by a drop door. Another door divides this subterranean cage from the steps leading to the surface of the garden.

X.—NOTES AT BANGALORE.

Hyder Ali, some time before his death in 1782, caused a large garden to be made and planted with mango trees, rather less than a mile east of Bangalore Fort. This is called the Lal Bagh and is to-day a very beautiful public garden about 100 acres (40·46 hectares) in extent, supported by the Government of Mysore and open free to the public daily.

Mr. G. H. Krumbiegel is the Superintendent and lives in a very nice official house in the garden.

Although principally a botanical garden, the Lal Bagh has for the last fifty years also contained a menagerie. The collection at present is not a large one, but I am told that plans are being prepared to extend it.

At the time of my visit, April 14, 15, and 19, 1913, this menagerie comprised :—

(i) The Court, probably built between 1850 and 1860, about 68 yards (62·17 metres) long by 26 yards (23·77 metres) wide, containing very solid dens for large carnivorous animals and a curious cage (now empty) formerly inhabited by a rhinoceros. A fine pair of tigers deserve notice, specially the female, said to have been presented by Mr. C. W. G. Morris in 1900.

(ii) The Aviary, now being rebuilt.

(iii) The Monkey House, the most noticeable inhabitant of which is a very large Mias or Orang-Utan, *Simia satyrus.*

(iv) The Great Paddock, which is so large that field glasses are necessary to see the animals in it, and contains, besides Blackbuck, Gazelles, Chital, and Sambur, a nice pair of Kákar, the Rib-faced or Barking Deer, *Cervulus muntjac.* One corner of this paddock is fenced in to form a separate enclosure for two Emus, *Dromaeus novaehollandiae.*

(v) The Bear House.

(vi) The Peacock Enclosure.

The Lal Bagh contains many really magnificent trees, including very large specimens of the Mango, *Mangifera indica*, the Mysore Fig, *Ficus mysorensis*, the Java Fig, *Ficus benjimini*, the Cunningham Fig, *Ficus cunninghami*, and *Spathodea companulata*. In India one sees from time to time interesting examples of the way in which various species of fig trees grow as parasites on trees of other genera; in the Lal Bagh I noted a *Ficus rhumphyi* growing in an Awla Tree, *Phyllanthus emblica*; and a Pipul, *Ficus religiosa*, growing in a *Feronia elephantum*. In the Mysore Zoological Garden were two Pipul Trees each growing in Neem Trees, *Melia azadirachta*, and at Quilon in Travancore I noted a *Ficus* of some species growing in a *Casuarina*.

There is a fine collection of bamboos, representing several genera and many species in the Lal Bagh, and in one of the tanks the pink flowered variety of the Sacred Lotus, *Nelumbium speciosum*, flourishes.

This lovely old garden, with its rich vegetation, makes of course a pleasant home for wild birds, which form one of its attractions; particularly noticeable in April were the Red-whiskered Bulbuls, *Otocompsa emeria subsp. incert.*, large numbers of Yellow Wagtails, *Motacilla flava subsp. incert.*, Bee-eaters, *Merops viridis*, Hoopoes, *Upupa indica*, and merry Parrakeets of the genus *Palaeornis*.

About a mile and a half north of the Lal Bagh, in the Cubbon Park, is the Bangalore Museum, also supported by the Mysore Government, and in charge of Mr. Krumbiegel. It is a fine building and contains an interesting zoological collection.

XI.—NOTES AT BARODA.

H.H. the Maharaja, the Gaekwar of Baroda, maintains between the city and cantonment of Baroda a large and beautiful garden, containing a collection of wild animals, which is open free to the public. The main feature of this garden is a deep nullah, with steep picturesque banks, which winds through the grounds and is crossed by several bridges.

Dozens of Langúr Monkeys, *Semnopithecus entellus*, roam wild in this garden, and I imagine that at times they must be a source of a good deal of trouble to the management. I visited the Baroda Zoological Garden on May 31 and June 1, 1913, and on the second of these days a large number of the staff seemed occupied in trying to get a langúr out of a deep dry pit into which it had fallen.

There is a " monkey tree " here, arranged on the same principle as the one in the Jaipur Ram-Newas Gardens, but with only three kennels, each occupied by a Bonnet Monkey, *Macacus sinicus* (*see* footnote on p. 96), and in a cage by itself in the Monkey House there is a large male monkey with pale yellow hair, the hair becoming almost golden yellow on its head, which appears to be a variety of this same species, *Macacus sinicus*, a very handsome variety that I do not remember ever having seen a specimen of before. The colour of its irides is normal brown. Except for its wonderful colour, it agreed with the Indian *Macacus sinicus*, and not with such few specimens as I have seen of the Ceylon Toque Monkey, *Macacus pileatus*. It may be mentioned that this remarkable animal was labelled " Yellow Baboon, Africa." *

There is a very fine open air enclosure for lions at Baroda, well over 240 feet (73·15 metres) in circumference. The iron fence surrounding it is of the same pattern as that of the lion and tiger enclosure in the Bombay Zoological Garden (*see* p. 48).

The most valuable animal in this collection was a Malay Tapir, *Tapirus indicus,* which was sharing a paddock with

---

* There was actually an African Yellow Baboon in the Baroda Zoological Garden.

some Nilgai, *Boselaphus tragocamelus,* an arrangement which I should be sorry to be responsible for, knowing the way in which Nilgai in captivity will at times make unprovoked and dangerous attacks on any living thing within their reach.*

The pretty little muntjac deer, which share a large aviary with rabbits, passerine birds, and pigeons, should also be mentioned.

The large "Flying Aviary" is very pretty and well planted; it contains specimens of twelve species of water-frequenting birds. In another aviary are some nice African Guineafowl of a species not often seen in captivity and that I do not know the name of.

A special house in this garden is called the Aquarium; it contains six wall tanks. These, at the time of my visit, were all empty.

The Baroda Museum is a large building in the public garden. On the staircase are busts of Linnaeus, Cuvier, Darwin, and Huxley. From some points of view this was the best Natural History Museum that I saw in India: it contains a very large general zoological collection but not much of *local* interest. Many of the exhibited animals are very well stuffed, and are the work of the well known London firm of taxidermists, Rowland Ward, Ltd. Some interesting models of Indian snakes should also be mentioned.

---

* I was told that a nilgai had lately actually killed one of the keepers in the Baroda Zoological Garden. Other accidents reported to me while in India, that I regret to have to record but which are useful to know of as warnings to men engaged with animals, are that of a male chital deer suddenly attacking and killing its keeper in the Peshawar Zoological Garden, a female sambar deer, with her fore feet, badly wounding u keeper in the Karachi Zoological Garden, the Reptile House keeper at Trivandrum being killed by a cobra, and the uufortunate man killed by a Hippopotamus at Calcutta.

XII.—NOTES AT BOMBAY.

Bombay Zoological Garden.

In November, 1862, a botanical garden was opened in the Byculla quarter of Bombay on the road to Parell, and during the seventies a small menagerie was built in the south-east corner of the grounds, in the place now occupied by the "Infirmary." In 1889 it was decided to make the institution a zoological as well as a botanical garden, and in 1890 many new cages were built and the collection of animals was increased.

This combined zoological and botanical garden is called the "Victoria Gardens," and is a municipal institution open free to the public, except on certain special days when a small entrance fee is charged.

The present Municipal Commissioner is Mr. P. R. Cadell, to whom I am indebted both for being given all the information I required about the finances of the gardens and also for copies of the plans of certain cages.

The Public Gardens Department of the Municipality includes the Victoria Gardens and eight smaller gardens in various parts of the city of Bombay.

The receipts from the Public Gardens go into the Municipal Treasury and for the year 1912–1913 amounted to 11,500 rupees (or about £767).

The expenses are provided for by the Municipality. The following table shows the amount of money, in rupees, available for expenditure for the year 1912–1913 :—

PUBLIC GARDENS.	MAINTENANCE.	NEW WORKS.	TOTAL.
	Rupees.	Rupees.	Rupees.
Budget Estimate for 1912–1913	87,345	11,798	99,143
Lapsed Grants for 1911–1912, renewed in 1912–1913	9,217	4,703	13,920
Additional Grants sanctioned in respect of unexpended balances in previous years ...	8,809	3,951	12,760
Additional Grants	1,000	—	1,000
Total Revised Estimate	106,371	20,452	126,823
Total Expenditure anticipated to be booked against Revised Estimate	96,096	16,881	*112,977

* Rupees 112,977 equal about £ 7,532.

The 1912–1913 Budget can be thus divided :—

Rupees.

(1) Victoria Gardens, salaries and upkeep		69,629
(2) ,, ,, music (band expenses)		3,500
(3) The smaller gardens, salaries and upkeep		7,316
(4) ,, ,, ,, music (band expenses) ...		6,900
(5) New works		11,798
TOTAL... ...		99,143

The 69,629 rupees available for the Victoria Gardens are thus appropriated :—

Rupees.

(1) Salaries :—
    (i) Employees on monthly pay 36,429
    (ii) Extra labour 1,000
                                         37,429

(2) Upkeep of grounds :—
    (i) Plants and seeds 1,000
    (ii) Tools, sand, manure, etc. ... 3,237
    (iii) Hydrants, maintenance of ... 600
    (iv) Roads ,, ,, ... 2,000
    (v) Drains ,, ,, ... 200
                                          7,037

(3) Upkeep of menagerie :—
    (i) Purchase of animals 3,000
    (ii) Travelling expenses 200
    (iii) Forage 12,500
    (iv) Medicines and disinfectants ... 250
                                          15,950

(4) Buildings and cages, repairs of	6,600
(5) Books, purchase and binding	250
(6) Printing	600
(7) Stationery	200
(8) Clock, maintenance of	100
(9) Ground rent	163
(10) Contingencies	1,300
TOTAL ...	69,629

I have arranged the above analyses differently from those published by the Municipality for the object of showing at a glance the disposition of the money in comparison with other gardens.

STAFF.

The staff of the Bombay Public Gardens on monthly pay numbers 206, distributed as follows :—

A. — VICTORIA GARDENS   Rupees per month.

1 Superintendent	450 and a house in Garden.
1 Assistant Superintendent	250 „ „ „
1 Overseer	80 „ „ „
4 Clerks	40 to 75
1 Label writer	40
1 Store-keeper	20
1 Head keeper of animals...	20
25 Keepers of Animals ...	11 to 16
1 Mahout for elephant ...	25
1 Cooly „ „ ...	12
1 Head gardener	25
1 Bouquet maker	20
1 Propagator	20
97 Gardeners	11 to 15
22 Women, garden sweepers	7
4 Carters	12
3 Messengers	12 to 13
13 Watchmen	12
4 Policemen	14 to 16

Total 183   3,104½   per month.

B. — THE SMALLER GARDENS.   Rupees per month.

18 Gardeners	11 to 13
3 Women	7
2 Policemen	13¾

Total 23   254½   per month.

The Superintendent, Mr. C. D. Mahaluxmivala, has general charge of the Victoria Gardens and the smaller gardens. The Assistant Superintendent, Mr. J. M. Doctor, C.M.Z.S., has special charge of the zoological collection in the Victoria Gardens.

The area of the Victoria Gardens is given by Murray ("Handbook to India," 8th Edition, 1911, p. 15) as 34 acres (13·75 hectares), but Mr. Mahaluxmivala informed me that the present area is about 50 acres (20·23 hectares).

Besides the money for new works granted from time to time by the Municipality, this Zoological Garden has been fortunate in receiving several donations for building its cages; thus in 1890–1891 the Bombay Tramway Company gave 10,000 rupees for new cages for bears and parrots; in 1891, Sir Dinshaw M. Petit, Bart., gave 1,500 rupees for an aviary; in 1894, H.H. Rasulkhanji Mohobatkhanji, Nawab of Junagadh, gave 2,500 rupees towards erecting a cage for the larger carnivora; and in 1899, the Maharaja Takhatsingji, G.C.S.I., of Bhownagar, gave 4,000 rupees for housing the smaller carnivora.

I visited the Victoria Gardens on April 12 and on June 3, 5, and 6, 1913, and was much impressed by their prettiness and by the large number of very nice trees, many of the trees being labelled with their names.

Large numbers of delightful little striped Palm-Squirrels, *Sciurus palmarum,* live wild in the gardens, and also many birds, the most noticeable species being the Indian House-Crow, *Corvus splendens,* the Magpie-Robin, *Copsichus saularis,* the Indian House-Sparrow, *Passer domesticus indicus,* and the White-breasted Water-hen, *Amaurornis phoenicurus.*

Noticeable features of these gardens are :—

(i) The Lake, one of the very prettiest bits of gardening I have ever seen. The scheme of planting is most effective. Some groups of Screw Pines, *Pandanus odoratissimus,* are particularly picturesque. One corner of this lake is enclosed by wire netting to form a " flying cage " and contains representatives of about twenty-five species of birds.

(ii) The Tebeldi, or Baobab Trees, *Adansonia digitata.* Although these Bombay specimens are small compared to the Tebeldis one is used to see in the Egyptian Sudan, nevertheless they are very nice trees. There is one on either side of the main path, just inside the entrance gate, and three others in the Lion Paddock.

(iii) The Stone Elephant, from the Island of Elephanta, near Bombay. The remains of this great statue are on the right of the entrance gate as one approaches the gardens.

(iv) The Lion Paddock, constructed in 1906 at a cost of about 9,000 rupees (£600), and the Tiger Paddock, constructed in 1909 for about 16,500 rupees (£1,100). It is difficult to believe that these really magnificient cages could have been built for such comparatively small sums.

The installation for lions consists of an enclosure about 290 feet (88·39 metres) in circumference, open at the top, and surrounded by an iron fence 20 feet (6·09 metres) in height; also a smaller enclosure about 110 feet (33·52 metres) in circumference, and two rooms, which the animals can be shut into : one, in interior dimensions, 18 feet (5·50 metres) in length by 12 feet (3·65 metres) in width, and about 10 feet (3·04 metres) in height; the other 10 feet (3·04 metres) by 11½ feet (3·50 metres) and of the same height as the larger room. Nine trees are growing in the Lion Paddock, which, as far as I could identify them, are :—

Three Tebeldis, *Adansonia digitata*, two "Custard Apples," *Polyalthia longifolia*, one Sissu, *Dalbergia sissoo*, one Palmyra Palm, *Borassus flabelliformis*, one *Garcinia sp.* and one *Sterculia sp.*

The installation for tigers consists of an enclosure about 340 feet (103·63 metres) in circumference, open at the top, and surrounded by an iron fence 20 feet (6·09 metres) in height, and three rooms, into which the animals can be shut. One room, in interior dimensions, 12 feet (3·65 metres) in length by 8 feet (2·43 metres) in width, and two rooms 9 feet (2·74 metres) by 8 feet (2·43 metres), all three rooms being 6 feet (1·82 metres) in height. Between these rooms and the open paddock there is a covered shelter for the animals, 33 feet (10·05 metres) long by 8 feet

(2·43 metres) wide and 6½ feet (1·98 metres) from floor to ceiling.

The iron fence, both for the lions and tigers, is curved inwards and downwards at the top, the points of the railings being about 3 feet 6 inches to 3 feet 9 inches (say 1·06 to 1·14 metres) inside the upright of the fence, and about 18½ feet (5·63 metres) from the ground level where the animals walk. The total height of the fence from the ground to the top of the curve is, as mentioned above, 20 feet (6·09 metres).

The collection of live animals represented, at the time of my visit, sixty-two forms of mammals, about eighty-three of birds, and three of reptiles. Among the more remarkable specimens were :—

(i) Two Long-haired Bunda Monkeys, *Macacus rhesus subsp.* (?)

(ii) A Brown Stump-tailed Monkey, *Macacus arctoides*.

(iii) A male Princess Beatrice's Antelope, *Oryx beatrix*, which has lived in the Bombay garden for eight and a half years. He carries a good pair of horns.

(iv) A male Beisa Antelope, *Oryx beisa*, from Jubaland.

(v) A Himalayan Black-throated Jay, perhaps *Garrulus lanceolatus*.

(vi) Three Giant Land Tortoises, *Testudo sp.*

Two of these are extremely large specimens. One, I was told, has been here about seventeen years (since 1896). These very valuable tortoises are kept in a paddock with black buck, adjutant stork, geese, etc. This appears to me to be running an unnecessary risk of the tortoises receiving an accidental injury from one or other of the other inmates of the paddock.

The Victoria and Albert Museum at the entrance of the gardens was commenced in 1862 and the building completed in 1871. It is also under the Municipality of Bombay but under separate management from the gardens. The Museum has a staff of nineteen men.

Bombay Crawford Market.

On the road from the Victoria Gardens to the Bombay Natural History Society's Museum is the Crawford Market, named after Mr. Arthur Crawford, who was Municipal Commissioner of Bombay from 1865 to 1871. From Murray ("Handbook to India," 8th Edition, 1911, p. 12) we learn that this market cost to build "over 11 lakhs of rupees," that is over £73,333.

On April 12, 1913, Mr. N. B. Kinnear and I visited this market. Among the numerous live animals for sale were monkeys of the genera *Cercopithecus* and *Macacus,* many Sacred Baboons, *Papio hamadryas,* some Marmosets, black and white Ruffed Lemurs, *Lemur varius,* specimens of the Red-fronted and of the White-fronted Lemur, *Lemur fulvus rufifrons* and *Lemur fulvus albifrons,* a large series of pet Cats and Dogs, several species of Myna-birds, also Minivets, Hill-Tits, White-eyes, Indian Sparrows, Buntings, Himalayan Goldfinches, Crested Larks, Short-toed Larks, a Roller, *Coracias indica,* large numbers of Parrots of Indian, Moluccan, and Australian species, Chukor Partridges, Jungle Fowl, Pheasants, Herring Gull, etc.

Gold and other fishes suitable for small aquaria can also be purchased here.

Bombay Natural History Society's Museum.

This most interesting Museum in Apollo Street, Bombay, is most deservedly of worldwide renown. The Journal published by the Society is recognized everywhere as one of the most useful and best illustrated of scientific journals, and in many ways this Society is carrying on such valuable work, as for instance the careful enquiry into the damage caused by termites ("white ants") and the means of preventing it, and also the very thorough survey of the mammals of India now in progress, that the house of Phipson and Company, in which the collections are lodged, may be looked upon as the zoological centre of the Indian Empire.

Mr. W. S. Millard, the Honorary Secretary, had left for Europe shortly before I arrived in India, but Mr. R. A. Spence, who was acting for him, and Mr. N. B. Kinnear, the Curator, were at the Museum when I visited it on April 11 and 12 and on June 2, 3, 4, 5, 6, and 7, 1913, and to these gentlemen I am very much indebted for their many acts of hospitality and for their great kindness in showing me specimens and books, in giving me introductions to people able to help me in my enquiries, and in giving me much useful information about India, means of communication, fauna and flora.

Besides the Museum and Library this Society maintains a small but select menagerie on its premises of such mammals, birds, reptiles, and fishes as are not too large for indoor life. Special attention may be drawn to the following :—

MAMMALS.

A female Malabar Squirrel, *Sciurus indicus,* a tame and friendly animal, now lives, attached by a chain, at the door of the Museum. A male of the same species died in March, 1901, after having lived over sixteen years in the Society's possession.

BIRDS.

The aviary contains specimens of the White-eye, *Zosterops sp.,* the Red-billed Hill-Tit, *Liothrix lutea;* three species of Bulbul, the Black-headed Myna, *Temenuchus pagodarum,* the Baya Weaver-bird, *Ploceus baya,* the Avadavat, *Sporaeginthus amandava,* the Rose-Finch, *Carpodacus erythrinus,* the Pale Rose-Finch, *Rhodospiza obsoleta,* the Black-headed Bunting, *Emberiza melanocephala,* the Indian Palm-Dove, *Turtur cambayensis,* the Rain-Quail, *Coturnix coromandelica,* which has been living here since 1907, and the Rock Bush-Quail, *Perdicula argunda.*

In the Secretary's room there is a Great Concave-casqued Hornbill, *Dichoceros bicornis,* presented by Mr. Ingle in August, 1894, so now eighteen years nine months and some days in captivity and still alive and vigorous.

Reptiles.

A specimen of the terrapin described by Dr. N. Annandale as *Geoemyda indopeninsularis,* which was obtained at Darwar and presented by Mr. G. C. Shortridge.

The Indian Chameleon, *Chamaeleon calcaratus.*

Two Indian Pythons, *Python molurus.*

A Long-nosed Tree-Snake, *Dryophis mycterizans.*

An Indian Cobra, *Naia tripudians.*

A Tree-Viper, *Trimeresurus gramineus.*

XIII.—NOTES AT CALCUTTA.

Calcutta Zoological Garden.

The Zoological Garden of Calcutta is situated in Alipore, just south of Tolly's Nulla. On the west the garden of the observatory adjoins it, and on the east Belvidere road divides the part of the zoological garden open to the public from the part reserved for administrative work.

The garden was founded in 1875 and opened to the public in 1876. Though a Government institution the Government of Bengal do not directly administer it, but manage it through an Honorary Committee, the members of which are appointed by Government. Lieutenant Colonel E. H. Brown, a retired officer of the Indian Medical Service, is the present Honorary Secretary of the Committee. Under the orders of this Committee is a paid staff, the principal executive officer being the Superintendent, who is provided with an official house in the garden. Another house, known as the "Hermitage," provides a residence for the Honorary Secretary, but the present holder of that office does not occupy it, but lives in the city of Calcutta.

The garden is financially supported by Government grants and by the gate-money, and has also from time to time received handsome donations of money from many Indian princes and other private individuals.

On most days the entrance fee is one anna (1*d.*, or four milliemes) per person, but on certain occasions this is raised to four annas (4*d.*, or 16 milliemes), or to one rupee (1*s.* 4*d.*, or 65 milliemes), and on a few days in the year the garden is open free to the public.

This institution is well known to zoologists, who have not personally visited Calcutta, by two useful publications :—
    (i) " Guide to the Calcutta Zoological Gardens," by the
        late Dr. John Anderson, F.R.S., 1883.

(ii) " Handbook of the Management of Animals in
   Captivity in Lower Bengal," by the late
   R. B. Sányál, 1892.

Rai Ram Bramba Sányál Bahadur, who died on October 13,
1908, will always be remembered in connection with the
Calcutta Zoological Garden, in which he worked for thirty-
three years.

Mr. Bijay Krishna Basu, Veterinary Inspector, was
appointed Assistant Superintendent of the garden on
February 25, 1907, and after the death of Mr. Sányál succeeded to the Superintendentship. Babu Somadev Ganguli
is the Assistant Superintendent.

Staff.

1 Superintendent.
1 Assistant Superintendent.
1 Overseer.
1 Store clerk.
1 Gate clerk.
58 Men (keepers, gardeners, etc.) on the permanent list.
35 Men (keepers, gardeners, etc.) on the temporary list.

Total 98.

Area.

The present area of land occupied by the Calcutta
Zoological Garden is about 53 acres (21·44 hectares) ; of
this, 33 acres (13·35 hectares) are open to the public on the
west side of Belvidere Road, and about 20 acres (8 09
hectares) lying on the east side of the road are used as qua-
rantine yard, stores and vegetable garden, and also give
room for quarters in which the keepers live. This year
(1913) the Government has allotted a further 11 acres
(4·45 hectares) for an extension of the zoological garden.
At the time of my visit this piece of land had not yet been
taken over, but when this is done the total area will be
about 64 acres (25·89 hectares).

Principal Features.

The principal feature of this garden is the Great Lake
(see Pl. II), a beautiful piece of water, surrounded by

magnificent trees and thronged by countless numbers of water-loving birds, the most abundant species being the Little Cormorant, *Phalacrocorax javanicus*, and the Indian Darter, or Snake-bird, *Plotus melanogaster*. Some Malay Tapirs, *Tapirus indicus*, have been turned loose in this lake ; it is found that they do no damage, either by frightening the birds or by injuring the vegetation, and the presence of these large, parti-coloured animals, add interest to the scene.

INSTALLATIONS FOR ANIMALS.

The principal buildings in which the animals are caged may be thus summarized :—

(1) *Primates.* — The "Dumraon House," named in honour of the Maharaja of Dumraon, built in 1878, and the "Gubbay House," named in honour of the late Mr. Elias Gubbay. Both these houses were remodelled in the years 1906–1907.

(2) *Carnivora.* — The "Burdwan House," built at the expense of the Burdwan Raj Estate, contains the lions and tigers, adjoining which is an open-air enclosure, the cost of which was met from the fund provided by the late Maharaja Bahadur Surya Kanta Acharya Chowdhury of Mymensing, surrounded by an iron fence curved inwards at the top, the height of the fence from the ground level where the animals walk to the top of the curve of the fence being apparently about 22 feet (6·70 metres).

The "Small Carnivora House," built in 1898, contains leopard, wild cat, hyaena, etc.

The "New Bear House," a present from the Nawab Bahadur of Dacca, and the "Abdul Ghani House," provide accommodation for the bears.

The "Mullick House" contains mongoose, otter, etc., and was named in honour of Raja Rajendra Nath Mullick of Chorebagan, Calcutta, who, I learn from Mr. Basu ("Guide to the Zoological Garden, Calcutta," 1910, p. 26), was an enthusiastic animal fancier and maintained his own private menagerie long before this zoological garden came into existence.

(3) *Ungulata.* — Besides the series of paddocks for cattle, antelope, deer, etc., there is a large enclosure for rhinoceros; another for hippopotamus, which is called the " Buckland Enclosure " to commemorate the name of Mr. C. T. Buckland, I.C.S., who was for many years President of the Garden; and the " Ezra House," built at the expense of the late Mr. David Ezra for the accommodation of the splendid pair of giraffes which he presented to the Garden in 1877,"* now occupied by zebras and wild donkeys.

(4) *Birds.*—The " Murshidabad House," a present from H.H. the Nawab Bahadur of Murshidabad, consists of a central hall and a series of very beautiful aviaries, in which banana and papya trees are growing, and contains such valuable exhibits as birds of paradise and toucans.

The " Sarnomoyi House," named in honour of the late Maharani Sarnomoyi of Cossimbazar, and erected at the cost of her nephew Maharaja Manindra Chandra Nandy, contains a vast collection, chiefly passerines, picarians, parrots, pigeons, and game birds.

The " Schwendler House," named after the late Mr. L. Schwendler, has pretty aviaries, principally for the smaller sorts of water-loving birds, teal, rails, etc.

The " Birds of Prey Aviary " is composed of a series of very fine lofty cages in which the vultures, eagles, and owls are lodged.

The " Smaller Duck House " contains flamingoes and other interesting birds, and the " Duck Pond," which is covered in with wire netting, contains a wonderful collection of anseres, herodiones, etc., rich in species and individuals. Many of the birds nest in this fine cage.

(5) *Reptiles.*—The " Crocodile Pool," built in 1878 as a pit with a rockery in it to keep snakes in, was adapted to its present use in 1907. It consists of two cages, one containing a large American alligator and the other three Indian crocodiles.

---

* B. Basu, " Guide to the Zoological Garden, Calcutta," 1910, page 39.

The "Reptile House," built in 1892, consists of a hall surrounded by cases and with two large tanks sunk in the centre of the floor. The visitors enter by a door in the east wall and pass round between the floor tanks and the wall cases.

ANIMALS.

I visited the Calcutta Zoological Garden on May 10, 11, 12, 13, and 14, 1913. The number of species seen is given in the "Analysis" on pages 8 to 11, but the following animals should be specially mentioned :—

Mammalia.

(1) *Primates.*

An Albino Bandar, or Bengal Monkey, *Macacus rhesus.*
Two Golden Haired Bandars, *Macacus rhesus var.*(?)
Three Brown Stump-tailed Monkeys, *Macacus arctoides.*
N.B.—The celebrated big Mandrill, *Papio maimon*, of the Calcutta Zoological Garden, died shortly before my visit. I was told it had lived here for twenty-nine years.

(2) *Carnivora.*

An Ounce, or Snow Leopard, *Felis uncia.*
A Clouded Tiger, or Clouded Leopard, *Felis nebulosa.*
A Crab-eating Mongoose, *Herpestes urva.*
Three Pandas, or Red Cat-Bears, *Aelurus fulgens.*

(3) *Ungulata.*

A magnificent pair of the Great One-horned Rhinoceros, *Rhinoceros unicornis.*

It is to be hoped that these very rare and valuable animals will breed here. As far as I know there is no record of a *Rhinoceros unicornis* having been born in captivity, but there seems no reason why such an event should not take place, as an Asiatic Two-horned Rhinoceros, *Rhinoceros sumatrensis*, was born in the Victoria Docks, London, on board the S.S. "Orchis," December 7, 1872 (*vide* Bartlett, P.Z.S., 1873, p. 104), and another of the same species in Calcutta in 1889 (*vide* Basu, "Guide to the Zoological Garden, Calcutta," 1910, p. 14).[*]

Four Malay Tapirs, *Tapirus indica.*

Four Gayals, *Bos frontalis.*

Two Bantings, *Bos sondaicus.*

An American Bison, *Bos bison.*

An Anoa, *Bos depressicornis.*

A pair of African Waterbuck, *Cobus unctuosus* (apparently).

A male Blessbok, *Damaliscus albifrons.*

A male Princess Beatrice's Antelope, *Oryx beatrix.*

A female Addax, *Addax nasomaculatus.*

Five Four-horned Antelopes, *Tetraceros quadricornis.*

Four Bárasingha, or Swamp Deer, *Cervus duvauceli.* At the time of my visit the big stag was making the garden resound with its voice, a loud "bray."

Fourteen Brow-antlered Deer, or Thameng, *Cervus eldi.*

Five Rib-faced, or Barking Deer, *Cervulus muntjac.*

A pair of Hippopotamus, *Hippopotamus amphibius.* Both the male and the female are very dangerous; I was told that about eighteen months ago the male killed his keeper, suddenly attacking and crushing the man in his mouth.

(4) *Marsupialia.*

Two Tasmanian Devils, *Sarcophilus ursinus.*

An albino Kangaroo, and an albino Wallaby.

AVES.

(1) *Passeres.*

Lesser Bird of Paradise, *Paradisea minor,* a cock and a hen.

Greater Bird of Paradise, *Paradisea apoda,* one cock.

Red Bird of Paradise, *Paradisea rubra,* four cocks.

Twelve-wired Bird of Paradise, *Seleucides nigricans,* one cock over seven years here.

I noted that the Rose-coloured Starlings, *Pastor roseus,* in the Calcutta and other Indian Zoological Gardens were in their beautiful pink plumage, which they do not assume in Giza or, so far as I have seen, in captivity in Europe.

---

[*] *Vide* Sányál "Handbook", 1892, pages 133, 134.

(2) *Herodiones*.

An Indian Open-bill Stork, *Anastomus oscitans*, a very rare bird to see alive and well in captivity. Mr. Basu told me that he had had it now for over a year.[*]

(3) *Limicolae*.

Two Great Stone-Plovers, *Esacus recurvirostris,* one of which was presented several years ago by Mr. Frank Finn. As Mr. Basu writes in his "Guide," page 44 : "No visitor can miss the stone-plover, which, with unfailing regularity, follows the footsteps of every passer-by in hopes of food or notice."

REPTILIA.

Two enormous Land Tortoises, probably *Testudo gigantea.*
Five Gharials, *Garialis gangeticus.*
Three Water Waran-Lizards, *Varanus salvator*, one of these being a particularly big specimen.

Calcutta Botanical Garden.

The "Royal Botanic Garden," as it is officially called, of Calcutta, is situated at Sibpur, on the west bank of the River Hugli, opposite Garden Reach.

This garden was "founded in 1786 on the suggestion of Colonel Kyd, who was appointed the first Superintendent" (Murray, "Handbook to India," 8th Edition, 1911, p. 65); it is 278 acres (112·49 hectares) in area.

This magnificent institution is too well known to require any description in this report. I will only mention the great banyan tree and give some details of the staff kindly given to me by Mr. G. T. Lane, who was so good as to show me over the gardens when I visited them on May 12, 1913, and to answer my many qustions on matters concerning the management of public gardens.

The celebrated Banyan Tree, *Ficus bengalensis*, is now

---

[*] In the "List of Animals in the Garden on 31st March, 1913," pages 36 to 46, of the "Report of the Honorary Committee for the Management of the Zoological Garden, Calcutta, 1912-13," this *Anastomus* does not appear to be mentioned.

about 144 years old; in October, 1908, the following information was officially given about it :—

"The circumference of its trunk, at $5\frac{1}{2}$ feet from the
ground, is 51 feet, and of its crown about 977 feet. Its
height is 85 feet. It has 562 aerial roots actually rooted
in the ground."

On the opposite side of India there was formerly an even
larger banyan tree, on an island in the Nerbudda, about
eleven miles from Broach. Forbes, who visited Broach
(1776–1783), says in his "Oriental Memoirs" (i, p. 26)
that this tree, called the "Kabir wad," enclosed a space within
its principal stems 2,000 feet in circumference. It had 350
large and 3,000 small trunks, and had been known to shelter
7,000 men. (Murray, "Handbook to India," 8th Edition,
1911, p. 119.)

STAFF.

Superintendent, Mr. C. C. Calder.
Curator,   Mr. G. T. Lane.

Office, Herbarium, and Library.—Not noted.

Gardens.
- 14 Gardeners.
- 8 Mowers.
- 2 Sirdars, headmen of coolies.
- 71 Garden coolies.
- 1 Rat-catcher.
- 1 Assistant mason.
- 2 Punkah coolies.
- 1 Cattle-man.
- 3 Ghoramies, employed in Palm House.
- 15 Sweepers.
- 12 Boys, employed in Nurseries.
- 11   ,,   ,,   in Orchid and Palm Houses.
- 1   ,,   ,,   watching waterfowl.
- 1   ,,   ,,   washing monuments.
- 1   ,,   ,,   in Stores.
- 58 Women, employed in sweeping roads.
- 22 Women, employed in Nurseries and Flower Garden.

Total 224

Fifteen men, one boy, and nine women are also on the
books, but were absent on the day the above enumeration was
made, making a total of 249 workers available in the garden.

In addition to this, the grass in the Palmetun is cut by women from Sibpur, and the clearing of water weeds from the lake is done by contract.

Calcutta Wild Fauna.

The Indian Fruit-Bat, or Flying Fox, *Pteropus medius*, is a very noticeable inhabitant of Calcutta. It is a curious and interesting sight to see these bats by day roosting in trees, hanging head downwards and appearing like some strange, large, brown fruit.

The pretty little striped Palm-Squirrels, *Sciurus palmarum*, are very numerous.

Of the many birds I saw in and about this city, besides the Crows, Babblers, King-Crows, Tailor-birds, Golden Orioles, Mynas, Magpie-Robins, Sparrows, Swallows, Coppersmiths, Rollers, Kingfishers, Swifts, Koels, Parrakeets, Vultures, Kites, Cormorants, and Pond Herons, I would like specially to mention the Indian Tree-Pie, *Dendrocitta rufa*, the White-breasted Water-hen, *Amaurornis phoenicurus*, the Indian Darter, or Snake-bird, *Plotus melanogaster*, the Night Heron, *Nycticorax griseus*, and also a flock of eight beautiful little Cotton Teal, *Nettopus coromandelianus*, which were on the Hugli close to the Botanical Garden.

Calcutta Museum.

The Indian Museum, an immense and magnificent building in the Chowringhee Road, Calcutta, is of such worldwide scientific reputation, and the collections of animals, antiquities, books, meteorites, minerals, etc., it contains so vast, that I cannot attempt to describe it. I was not fortunate enough to meet Dr. N. Annandale, the Superintendent, as he was at Simla when I visited his Museum, May 13, 1913, but his assistants, Mr. S. W. Kemp and Mr. F. H. Gravely, very kindly conducted me round the galleries and laboratories, and Mr. G. E. Pilgrim, of the Geological Survey, was so good as to show me the wonderful series of Tertiary mammal remains from various parts of India and from the Island of Perim that are his special charge.

The elephant and crocodile specimens in this Museum are referred to in the chapters in this report devoted to those animals.

The undermentioned specimens in the lofty and well lighted exhibition galleries should be specially noted :—

(i) Skull and horns of a Kudu Antelope, *Strepsiceros sp. incert.*, brought from Abyssinia by the late Mr. W. T. Blanford, F.R.S.

(ii) A stuffed male Duyong, *Halicore dugong*, from the Gulf of Manar. *

(iii) A model of the Súsú, or Gangetic Dolplin, *Platanista gangetica.*

(iv) Models, made by C. Subraya Mudaliar of the Trivandrum Museum, of Little Indian Porpoise, *Phocaena phocaenoides*, Speckled Dolphin, *Sotalia lentiginosa*, and the Dolphin, *Delphinus delphis.*

(v) A model of the Soft-Turtle, *Chitra indica.*

(vi) A Rat Snake, or Dhaman, *Zamenis mucosus*, 8 foot 3 inches (2·51 metres) long, labelled as being the record for length; but Wall (J.B.N.H.S., XVII, p. 260) mentions specimens 9 feet 1½ inches (2·78 metres) and 11 feet 9 inches (3·58 metres) long.

(vii) The saw of a Saw-Fish, *Pristis perrottetii.*

The fish is said to have been 21 feet 7 inches (6·57 metres) in length. Indian Saw-Fishes of this great size are very seldom obtained. Day ("Fauna of British India," Fishes, I, 1889, pp. 38, 39) mentions a *Pristis cuspidatus* of 20 feet (6·09 metres) and a *Pristis pectinatus* of 24 feet (7·31 metres).

(viii) A specimen of the bright coloured Saw-Fish, *Pristis pectinatus annandalei*, from the coast of Burma.

(ix) A model of a green coloured Sting Ray, *Trygon sephen* (or *Hypolophus sephen*).

(x) A stuffed specimen of the very thorny Ray called *Urogymnus asperrimus.*

---

* Total length, as stuffed, from tip of snout along curves, 9 feet 4 inches (2·84 metres). Mr. S. W. Kemp very kindly had this specimen measured at my request.

XIV.—NOTES AT JAIPUR.

The Zoological Garden of Jaipur is in the "Ram-Newas Gardens," outside the city wall and open free to the public. According to Murray ("Handbook to India," 8th edition, 1911, page 141) this garden is 36 acres (14·56 hectares) in extent and was laid out by Dr. de Fabeck at a cost of 400,000 rupees, or £26,666.

I visited Jaipur on May 16 and 17, 1913 ; the principal features of this garden are :—

(i) The Great Paddock.

An enormous undulating piece of ground with a nulla running through it. There is nothing to compare with it in size in any zoological garden I know, except some of the enclosures for the Duke of Bedford's animals at Woburn Abbey in England. It is inhabited by a herd of Blackbuck, *Antilope cervicapra*, and by Sambar Deer, *Cervus unicolor*, and is resplendent with wild Peacocks, *Pavo cristatus*.

(ii) The Flying Cage.

A large cage, with much rockwork and water, containing the following birds, apparently all living together in harmony :—

4 Purple Coots, *Porphyrio sp.*

1 White Pelican, *Pelecanus onocrotalus.*

1 Dalmatian Pelican, *Pelecanus crispus.*

17, or more, Cormorants, *Phalacrocorax carbo.*

3 Indian Darters, *Plotus melanogaster.*

5 Black-headed Ibises, *Ibis melanocephala.*

9, or more, White Storks, *Ciconia alba.*

2 White-necked Storks, *Dissora episcopus.*

2 Indian Yabiru Storks, *Xenorhynchus asiaticus.*

17, or 18, Painted Storks, *Pseudotantalus leucocephalus.*

1 Grey Heron, *Ardea cinerea.*

2 Eastern Cattle Egrets, *Bululcus coromandus.*

(iii) The Large Waders' Aviary.

A very fine cage enclosing a large pond and containing over thirty-one Flamingoes, *Phoenicopterus roseus*, over twenty-four Barred-headed Geese, *Anser indicus*, other Geese, about twelve species of Ducks, mostly represented by dozens of individuals, and some Grey Cranes, *Grus communis*, and Demoiselles, *Anthropoides virgo*.

(iv) The Small Waders' Aviary.

A pretty cage inhabited by about fifty birds, Coots, Moorhens, and Plovers.

(v) The Collection of Passerine Birds.

This is far larger than any other I saw in India; about fifty species are represented and the individual birds can only be reckoned by hundreds. It is particularly rich in what are called "Soft-billed Birds," and is also of particular interest because it contains so many local birds.

The keepers in charge of the birds are Mohammedans.

(vi) The Monkey Tree.

A Banyan Tree fitted up as a home for captive monkeys. In the branches of the tree are wooden kennels, one for each monkey. From close to each kennel a perpendicular iron rod descends to the ground. Each monkey is chained to a ring. These rings slide on the perpendicular rods. Thus by means of these rods the monkeys can come and go as they please from the kennels and branches to the ground. There is also a somewhat similar "Monkey Tree" in the Baroda garden. The species kept in this tree at Jaipur are Bengal Monkeys, *Macacus rhesus*, and young individuals of the Sacred Baboon, *Papio hamadryas*.

(vii) The Wild Donkeys.

A very fine pair in one paddock. Both the stallion and the mare are big animals, in colour nearly white, with pale reddish brown patches on face, neck, and flanks, and a dark brown vertebral line which gets broader posteriorly. On both fore and hind limbs there are traces of stripes. According to Blanford ("Fauna of British India," Mammalia, 1891, p. 470) this would be a form of *Equus hemionus*, but according to Lydekker ("Guide to the Horse Family," Brit. Mus., 1907, p. 32) it would be *Equus onager typicus* of Northern Persia. I was unable to ascertain where these particular animals came from, but it appeared to be more likely that they were of Indian origin than imported from Northern Persia, especially as Blanford (*loc. cit. supra*) records that wild asses occur in Bickaneer, a neighbouring State to Jaipur in Rajputana.

In another paddock were two more of these fine animals, making four in all in the Jaipur collection.

Altogether I saw fourteen Asiatic Wild Donkeys in Indian Zoological Gardens. Two at Bombay were said to be from Kathiawar; Calcutta had two labelled *Equus onager*; Karachi, one said to be from Persia; Lahore two, very like those seen at Jaipur; Mysore a pair, said to be from Cutch; and Peshawar one, without locality.

In the Ram-Newas Gardens is also the Museum, called the "Albert Hall"; a very ornamental building designed by Colonel Sir Swinton Jacob, K.C.I.E. The zoological collection is unimportant.

Besides the collection of animals in the Ram-Newas Gardens there is also an official menagerie inside the City of Jaipur. This contained five tigers and two leopards; housed in a row of stone cages, with fronts of iron bars, and small retiring rooms at the back with iron-grated windows. The cage floors are of stone, but wooden boards have been provided for the beasts to lie on. The keeper in charge is a Mohammedan.

Some notes on the Mugger Tank at Jaipur will be found in this report in the chapter devoted to crocodiles.

No account of Jaipur can be in any way complete without some mention of the wild fauna of the place. Thanks to the protection wisely given to both bird and beast by H.H. the Maharaja and his subjects, the wild things here show wonderful confidence in mankind.

I saw great grey Langur Monkeys, *Semnopithecus entellus*, playing in the Palace Garden in the city, and a few miles away, when going up the Galta Pass in the hills, I came on a party of between thirty-two and forty of these animals. They sat in groups on the path and did not move to let me by ; in fact I had to move to right and left not to run into them, and I had to step carefully to avoid treading on their long tails.

Palm-Squirrels, *Sciurus palmarum*, abound in Jaipur as in so many other parts of India, and in the environs of the town there are many Blackbuck, *Antilope cervicapra;* in one herd of these graceful antelopes I counted over forty individuals.

Of the birds the most noticeable are the Peafowl, *Pavo cristatus*, which appear to be everywhere, and the Blue Rock Pigeons, *Columba livia* or *Columba intermedia*, which in untold thousands swarm in the streets and cover the roofs of the buildings. The gardens are full of Spotted Doves, *Turtur suratensis*, Palm-Doves, *Turtur cambayensis*, and Laughing Doves *Turtur risorius*. Crows, King-Crows, Sparrows, Bee-eaters, Hoopoes, Parrakeets, Vultures, and Kites abound. Quantities of Myna-birds, *Acridotheres tristis* and *Temenuchus pagodarum*, frequent the zoological garden, and at the Mugger tank I saw a few Pond-Herons, *Ardeola grayi*, and Black-winged Stilt Plovers, *Himantopus candidus*.

XV.—NOTES AT KARACHI.

The Zoological Garden of Karachi belongs to the Municipality of Karachi and is supported by grants given by that body. The garden contributes to the municipal funds by selling plants and fruit; one section of the grounds is devoted to a vinery, the grapes from which sell for about £100 a year.

The area of the garden is 46 acres (18·61 hectares); it is open free to pedestrians, and at a charge of 2 annas (2d. or 8 milliemes) for equestrians or visitors in carriages. The garden is, however, closed to the public for some hours in the middle of the day, say from noon to 3 p.m.

The Municipality delegates the administration to an Honorary Committee, the present chairman of this committee being Mr. H. P. Farrell, Principal of the Dayaram Jethmal Sind Arts College, Karachi. The superior paid officials are the Superintendent, Mr. Ali Murad, who occupies an official house in the garden, and the Overseer, Mr. L. P. D'Souza. To these gentlemen I am indebted for much useful information about their garden and the methods employed in managing and feeding their animals.

I visited the Karachi Zoological Garden on May 25, 26, and 27, 1913. The following notes may prove of interest :—

    (i) Langur Monkeys, *Semnopithecus entellus*, have bred in captivity here.

    (ii) A Bandar, or Bengal Monkey, *Macacus rhesus*, is said to have lived nearly twenty years here.

    (iii) The pair of Tigers, *Felis tigris*, have both been here since 1901.

    (iv) Two Jungle Cats, *Felis chaus subsp. incert.*, both caught in Karachi. These big cats come into the Karachi Zoological Gardens at night and are most destructive. Within the last few years, I am told, they have caused the death of a large number of birds and of a Mouse Deer, *Tragulus*.

(v) The otters are accommodated in a nice pond with an island and much vegetation.

(vi) A female Princess Beatrice's Antelope, *Oryx beatrix.*

(vii) A Goral, *Cemas goral.*

(viii) Four individuals of the Sind race of the Persian Wild Goat, *Capra hircus blythi* (*vide* Lydekker, "Brit. Mus. Cat.," Ungulate Mammals, I, 1913, p. 159). This fine species breeds here.

(ix) Six individuals of the Ghud, the Sind race of the Urial Wild Sheep, which may perhaps be identical with the Afghanistan race, *Ovis vignei cycloceros* (*vide* Lydekker, "Brit. Mus. Cat.," Ungulate Mammals, I, 1913, p. 89).

(x) Nine individuals of the Sind race of the Wild Boar, which appears to be a rather small form of *Sus cristatus.*

(xi) Two Eagle-Owls, which Mr. Farrell told me were of the Indian species *Bubo coromandus,* which have now been between five and six years here and have been fed only on butcher's meat, varied by an occasional dead Crow, *Corvus splendens.* These fine Owls appeared to be in excellent condition.

(xii) Eight grey, white-throated individuals of the very graceful Indian Reef-Heron, *Lepterodius asha.* A species of bird very seldom seen alive in a zoological garden.

Of the wild fauna of the Karachi Zoological Garden it may be mentioned that pretty little striped Palm-Squirrels, *Sciurus palmarum,* abound, that the Crows, *Corvus splendens,* are so numerous and prove themselves such a nuisance that they have to be systematically shot down from time to time, and that a few Cormorants and a large number of Pond-Herons and Night-Herons frequent the waterfowl ponds of their own accord.

The Victoria Museum, about a mile and a half from the Zoological Gardens, is open free to the public and kept up at the expense of the Municipality of Karachi. It is a museum of miscellaneous objects ; among the zoological collection attention may be called to the fine heads and horns of Sind Wild Goats and of Sind Wild Sheep, and to the large series of small local animals, preserved in spirit, which should be very useful material for any student of the Indian fauna.

Some notes on the Mugger Tank, ten miles from Karachi, will be found in this report in the chapter devoted to crocodiles.

XVI.—NOTES AT LAHORE.

The Lahore Zoological Garden occupies the west end of the "Lawrence Gardens," which cover an area of about 112 acres (45·32 hectares).

This zoological garden is under the Punjaub Government, and is supported by a Government grant and by a subsidy from the Lahore Municipality; it is in the charge of a paid Curator, Mr. A. W. Pinto, and is open free to the public.

The menagerie may be divided into three sections :—

(1) An inner part with the office, cages for primates and carnivora, aviaries, a snake house (erected November 15, 1912, but now empty), and some tanks containing many very big gold fish.

(2) A pretty shady pond for waterfowl, frequented by wild Night Herons, *Nycticorax griseus.*

(3) An outer part occupied by a series of large grass paddocks for ungulates and domestic poultry, and a paddock containing a pond and much vegetation, surrounded by a high wire fence, inhabited by a beautiful specimen of the Frontier Wolf, *Canis lupus* (*see* Pl. V).

This garden is celebrated for having formerly been the home of the tiger "Moti." The late Mr. J. L. Kipling, C.I.E., records two of the adventures of "Moti" in his book "Beast and Man in India" (1904), on pages 356 and 357.

When I visited Lahore on May 20 and 21, 1913, there were no tigers living in the collection, which, however, contained a Tahr, *Hemitragus jemlaicus,* a species of Indian wild goat frequently seen in zoological gardens in Europe, but the only specimen I saw in India. The Monál or Impeyan Pheasant, *Lophophorus refulgens*, should also be specially mentioned. Pretty little striped Palm Squirrels, *Sciurus palmarum*, abound here, and among the wild birds were the Punjaub Bulbul, *Molpastes intermedius*, the White-eared Bulbul, *Molpastes leucotis*, the Black Robin, *Thamnobia cambaiensis*, the Magpie-Robin, *Copsychus saularis*, and the Indian Palm-Dove, *Turtur cambayensis*.

The Museum at Lahore, called the "Jubilee Museum," though chiefly devoted to antiquities and art, contains a small zoological collection, including some fine horns and antlers of Northern Indian big game.

Among the exhibited series of weapons there is a "Fish Sword," the saw of a Saw-Fish, *Pristis sp.*, with the basal end cut into a hilt, capable of being grasped in the hand.*

This Museum owes a great deal to the work of the late Mr. J. L. Kipling, C.I.E., and Mr. Rudyard Kipling has acted as its Curator.

---

* In the Museum at Bristol, England, in a case marked "Solomon Islands," there is a somewhat similar "Fish Sword," also made from the saw of some species of *Pristis*. The grip is improved by pieces of shagreen inserted into the carved out base of the saw.

XVII.—NOTES AT MADRAS.

Madras Zoological Garden.

A Municipal Zoological Garden occupying one end of the People's Park. This park is 116 acres (46·94 hectares) in area and is open free to the public ; the Zoological Garden itself is enclosed by a corrugated iron fence and an entrance fee of ½ anna (½d. or 2 milliemes) is charged per visitor.

Messrs. Higginbotham & Co. of Madras published " A Guide to the People's Park, Madras," in 1876.

The site of the Zoological Garden is flat and occupies three sides of a lake, on which is an island connected to the mainland by two wooden bridges. Pelicans, adjutant storks, white swans, black swans, and peafowl are loose in the grounds during the day, but these birds have to be shut up in cages every night for fear of the jackals which roam about the city.

I visited the Madras Zoological Garden on April 30 and May 1, 1913, and have to express my best thanks to Mr. P. L. Moore, C.I.E., I.C.S., the Municipal Commissioner, and to Mr. H. Garwood, the Superintendent of the Park, for their kindness in showing me round their garden and giving me useful information about it.

Except the Tiger House, which was built in 1887, the existing cages for the larger carnivora at Madras date from the year 1858. Mr. Moore told me that larger and better quarters are about to be built for these animals.

Many pretty little striped Palm-Squirrels, *Sciurus palmarum*, occur wild in this garden, and Crows, both *Corvus splendens* and *Corvus macrorhynchus*, abound. The *Corvus splendens* gives a lot of trouble by perching on the backs of and worrying some of the animals, especially the Sambar deer and the Lamas.

Special attention may be called to the following animals in the Madras Garden :—

(i) A very large and dark coloured male Mias, or Orang-Utan, *Simia satyrus*.

(ii) A female Asiatic Two-horned Rhinoceros, *Rhinoceros sumatrensis*, which has now been about fourteen years here.

(iii) A male Malay Tapir, *Tapirus indicus*, which has a nice cage to live in.

(iv) Two Great Black-headed Gulls, *Larus ichthyaëtus*, a very handsome species, rarely seen in menageries.

(v) Six Brown-headed Gulls, *Larus brunneicephalus*.

(vi) An Indian Python, *Python molurus*, which was purchased on August 3, 1903, and is now in the most beautiful condition. It appears to be not only the largest specimen of this species but the most magnificent individual snake that I have ever seen alive.

Madras Horticultural Garden.

I visited this very nice, neat, pretty and instructive garden on May 1, 1913. Though not fortunate enough to meet any of the officials in charge, I learn from Murray ("Handbook to India," 8th Edition, 1911, p. 408) that this garden was founded about 1836, "mainly through the efforts of Dr. Wright," and that it occupies an area of 22 acres (8·90 hectares).

Among the many interesting plants I can only mention here the Bread-Fruit Tree, *Artocarpus incisa*, laden with quantities of fruit, and fine specimens of the Sausage Tree, *Kigelia pinnata*, and of the Mahogany, *Swietenia mahagoni*; and among the delightful wild birds which were living in this quiet beautiful place a Crow-Pheasant, *Centropus sinensis*, in a clump of small palms, *Nipa fructitans*, and a Black Bittern, *Dupetor flavicollis*, sitting motionless on the edge of a pretty pond.

Madras Museum.

A great institution under the superintendentship of Dr. J. R. Henderson, comprising the Connemara Library and large and valuable collections of arms, antiquities, industrial arts, and natural history. Among the many interesting specimens in the zoological section attention may be called to a skull with very fine horns of a wild Indian Buffalo, *Bos bubalus*, to a stuffed female Duyong, *Halicore dugong*, from Tuticorin, 6 feet 9½ inches (2·07 metres) long,* and to the skin of a Sea Snake, labelled as *Distira spiralis,* said to be 8 feet 2 inches (2·48 metres) long.†

In the hall of the Madras Museum there is a collection of live animals in cages, comprising gerbils, owl, pigeons, jungle fowl, tortoises (*Testudo elegans* and *Testudo travancorica*), lizards, snakes, fishes, and scorpions. The collection of snakes is particularly representative and instructive, including, on the day of my visit, May 1, 1913, specimens of :—

1. The Indian Python, *Python molurus.*
2. The Conical Sand-Boa, *Gongylophis conicus.*
3. The Indian Sand-Boa, *Eryx johnii.*
4. The Painted Tree-Snake, *Dendrophis pictus.*
5. The Long-nosed Tree-Snake, *Dryophis mycterizans.*
6. The Cerberus Water-Snake, *Cerberus rhyncops.*
7. The Krait, *Bungarus caeruleus.*
8. The Indian Cobra, *Naia tripudians.*
9. The Russell Viper, *Vipera russellii.*
10. The Carpet Viper, *Echis carinata.*

Madras Aquarium.

I visited this very attractive aquarium on April 30, 1913. Dr. J. R. Henderson, of the Madras Museum, is the Director, and Mr. Pareasureamean the keeper in charge of the collection. Besides some marine turtles, I saw representatives of about fifty-four species of fishes. The following list gives a comparative idea of the size of the Madras

---

* Dr. J. R. Henderson very kindly had this specimen measured for me.

† Boulenger ("Fauna of British India," Reptilia, 1910, p. 401), writes of *Hydrophis spiralis :* "Total length 6 feet (Günther). I have only seen young specimens." A Sea Snake, probably *Distira brugmansi,* 9 feet long, from Penang, has been lately recorded. *Vide* Stone, "Journal Bombay Nat. His. Soc.," XXII, Sept. 1913, p. 403.

collection of fish compared to other aquariums that I have
visited during the last six years :—

Frankfort-on-Main ... ... 85 to 91	forms of fish seen.
Amsterdam 80	,, ,,
Prighton 54	,, ,,
Madras. 54	,, ,,
Berlin (now closed)... 52	,, ,,
Naples ... ... 40 to 47	,, ,,
Stibbington Hall (Capt. Vipan's) ... 44	,, ,,
Blackpool ... ... 23 to 34	,, ,,
Hamburg ... ... about 30	,, ,,
Gezira, Cairo ... ... 27 to 30	,, ,,
Trieste ... ... about 26	,, ,,
Liverpool 21	,, ,,
Crystal Palace, London 18	,, ,,

Dr. Henderson has written an interesting "Guide to the
Marine Aquarium," published in 1912 by the Madras
Government, Madras, from which I quote the following
extracts :—

" The Marine Aquarium, which is the first institution of
its kind in India, was opened to the public on 21st October,
1909. It was erected by the Madras Government with the
two-fold object of providing an attractive display of living
fish, and of furnishing the material for a scientific study of
the fish and other marine animals of the Madras coast.

" The main entrance leads into a paved area with a central
fresh-water pond, and on either side are placed five large
tanks with plate-glass fronts. The seaward side of the
central area is occupied by a large open tank stocked with
turtles. At convenient places there are small isolated aquaria
for special novelties, which for various reasons cannot be
exhibited in the larger tanks, and with the exception of one
or two of these, which contain fresh-water fish, and also of
the central pond, all the tanks contain sea water.

" Sea water is conveyed to a covered well in the rear of the
Aquarium, along a pipe filled by hand at the seaward end.
From the well it is pumped into filter-beds, and from these
passes to large elevated cisterns, whence it is distributed to
the tanks. Fish in common with other living animals require
a plentiful supply of oxygen, and this is ensured partly by
direct aeration and partly by a circulation of water from

tank to tank. The method of aeration consists in forcing air by means of a hand pump into compression cylinders, whence it is distributed along slender tubes to the different tanks. Each distribution tube is connected with a filter candle through which the air is driven into the tank in a cloud of small bubbles. The air pump is worked at intervals, both by day and by night, for experience has shown that if the supply of air were cut off for a comparatively short time all the inhabitants of the tanks would perish. The fish are living permanently under conditions of overcrowding which could only be found temporarily in the open sea. Similarly, in the case of a building crowded with human beings, a proper supply of fresh air must be admitted or the people will suffer. The water which has circulated through the tanks can, if desired, be brought back to the filter-beds, and, after purification, used a second time.

"Nearly all the exhibited specimens have been captured within a few miles of the Aquarium. The fish are caught in ordinary fishing nets, many of them in the sheltered waters of Madras harbour, but in the absence of steam trawlers, or of boats fitted with tanks, considerable difficulty is experienced in bringing them alive to the Aquarium. Indeed, for every fish safely placed in a tank a great many die in transit, as a result of accident, exposure to the sun, or want of air. Curiously enough, one of the commonest causes of fatality exists in injuries to the eyes. The fish, on being removed from their natural element, often dash wildly against the sides of the tub or other vessel in which they are temporarily placed, with the result that their eyes, which are unprotected by eyelids, frequently get damaged. Once the fish are fairly established in their new quarters the chief difficulty has been overcome, and, accidents excepted, they are likely to live a reasonable period of time. At the time of writing (March, 1912) there are several fishes in the Aquarium which have been in residence for over two years.

" The fish and turtles are fed once daily, their food consisting chiefly of chopped-up fish, crustaceans such as a crab found commonly in the sand within the surf zone (*Hippa asiatica*, the " gilly poochi "), and sand-worms.

" *Invertebrate Animals.*—For various reasons comparatively few of these are exhibited in the Aquarium. In the first place it is difficult to keep them in the same tanks with fish, to the rapacity of which they soon fall victims, and secondly, none of them thrive particularly well under the artificial conditions to which they are subjected. In the small floor aquaria will, from time to time, be found cuttle-fish, hermit-crabs, sea-anemones, and other invertebrates which are described in special labels. Certain of the larger crustacea are occasionally exhibited in tank 2, and of these some beautiful forms occur on the Madras coast, notably the large lobster (*Panulirus*), and brightly coloured swimming-crabs (*Neptunus*). The South Indian sea-anemones, shown in floor aquarium E, do not compare favourably with the large showy specimens which form such an attractive display in European aquaria.

" *Fresh Water Animals.*—The place taken by these in the Aquarium is an altogether secondary one, and, as already remarked, all the large tanks contain salt water. In spite of this fact there will be seen in tank 3 a number of specimens of the perch-like fish, *Etroplus suratensis*, where they have lived and flourished for upwards of two years, and there are other local fish, which appear to live equally well in fresh or in salt water. In the fresh-water tanks aeration is secured by growing pond weeds, which during the daytime liberate a sufficient quantity of oxygen to meet the requirements of the animal inmates.

" On account of curious incompatibilities of temper, it has been found impossible to arrange the bony fish according to any scientific classification, for in several instances experience has shown that even closely related species will not dwell amicably together. Some specimens exhibit such a pugnacious disposition that they have to be removed from the Aquarium altogether. In a few cases, for the above reason, single individuals have had to be removed from the companionship of the other members of their species, and placed in a separate tank. It would of course be possible to obviate this were the number of exhibited species strictly limited, but by so doing it is felt that a good deal of the attractive-ness furnished by a large variety of fish would also be lost.

The coloured illustrations placed above the waterline of the various tanks represent common or peculiar Madras fish, labelled with their scientific and Tamil names, and all of them have at some time or other been included in the collection. For obvious reasons it is impossible to ensure that all the fish figured will be present in the Aquarium at any given time."

Of the fish that I saw alive in the Madras Aquarium attention may be specially called to the following :—

Family Notidanidae.

(i) The very curious Shark, *Notidanus indicus.*

Family Scyllidae.

(ii) Seven specimens of the Dog-fish, *Chiloscyllium indicum.*

Family Muraenidae.

(iii) Eels of the species *Muraena punctata* and *Muraena pseudothyrsoidea.*

Family Percidae.

(iv) A fish, very like the "Ishr" of the Nile, called *Lates calcarifer·* The late Dr. F. Day ("Fauna of British India," Fishes, I, 1889, p. 441) mentions that this species reached 5 feet (1·52 metres) in length and 200 lbs. (90·71 kilogrammes) in weight.

(v) Another species of Sea-Perch, *Lutjanus jahngarah.*

(vi) Yet another Sea-Perch, *Therapon jarbua.*

Family Squammipinnes (or Chaetodontidae).

(vii) Six specimens of *Heniochus macrolepidotus,* which, as Dr. Henderson says, is a fish of striking coloration, with its black bands and yellow fins. The dorsal fin carries a long white streamer. The keeper told me that some individuals have lived in this Aquarium for four years.

Family Cirrhitidae.

(viii) The Sangan, *Cirrhitichthys aureus.*

Family Scorpaenidae.
(ix) Six specimens of the Scorpion-Fish, *Pterois russellii.*

Family Berycidae.
(x) The Mundakankākāsi, *Muripristis botche.*

Family Carangidae.
(xi) A kind of " Horse-Mackerel," *Caranx leptolepis.*

Family Scombridae.
(xii) The Sucker-Fish, *Echeneis naucrates.*

Family Chromides.
(xiii) *Etroplus suratensis* (mentioned above, page 77) which in shape resembles the " Bulti " of Egypt.

Perhaps the most fascinating of all the wonderful things to be seen in the Madras Aquarium were the representatives of the Plectognathi :—

(xiv) Actually five different species of File-Fish or Trigger-Fish, *Balistes.* Some individuals had lived here for several years, according to the keeper.
(xv) Two specimens of the very curious Coffer-Fish, *Ostracion cornufer.* The pleasure of seeing these extraordinary and charming little fishes alive and apparently well and happy was alone worth any discomfort that a visit to India during the hot season might entail.
(xvi) Representatives of two species of Globe-Fish, *Tetrodon.*

XVIII.--NOTES AT MYSORE.

The Sri Chamarajendra Zoological Garden in Mysore, which is named after its founder the late Maharaja of Mysore, was started in December, 1892, but I am told the buildings at present existing have been practically all constructed since 1909.

The present area of the Zoological Garden is about 15 acres (6·07 hectares). Another 20 acres (8·09 hectares) are available for future extension, making a total of 35 acres (14·16 hectares). A Zoological Museum is to be built on part of this land.

The whole institution is the private property of H.H. the Maharaja of Mysore and kept up at his own expense. The public are admitted to the garden at the nominal fee of one anna (1d., or 4 milliemes) each person.

I visited this very nice neat and clean garden on April 16, 17, and 18, 1913, and am very much indebted to Mr. R. H. Campbell, I.C.S., Private Secretary to H.H. the Maharaja, to Mr. Mirza M. Ismail, H.H.'s Huzur Secretary, and to Mr. A. C. Hughes, the Superintendent of the Zoological Garden, for their kindness in showing me the grounds and collections and giving me much useful and very interesting information about Mysore and its fauna.

STAFF.

The staff of the Mysore Zoological Garden consists of :—

	1	Superintendent.
	1	Assistant to the Superintendent.
	1	Clerk.
	2	Gate keepers.
Menagerie	1	Head keeper.
	13	Keepers.
Gardens	8	Gardeners on Permanent List.
	11	Gardeners on Temporary List.
	10	Convicts deputed from the State prison to help in the garden.
Total	48	

The elephants and camels which H.H. the Maharaja sends to the garden for visitors to ride on are attended by their own mahouts and camel-men, who are not keepers on the establishment of the Zoological Garden and are not included in the above list.

A feature of the Mysore Zoological Garden, in which it differs from all the other zoological gardens I saw in India, is that it stands on high ground, and from it magnificent views can be obtained of the neighbouring country, particularly of the grand hill called Chamundi on the far side of the river valley. Other features of this garden to be specially noted are:—

(i) The ponds with the pink flowered variety of the Sacred Lotus, *Nelumbium speciosum.*

(ii) The fine collection of Primates, including an albino individual of the Macaque, or Crab-eating Monkey, *Macacus cynomolgus*, three Mandrills, *Papio maimon*, two Drills, *Papio leucophœus*, and two very interesting specimens of the local race of the Slender Loris, *Loris gracilis.**

(iii) A magnificent pair of Tigers, *Felis tigris*, from Mysore.

(iv) Three very large Leopards, *Felis pardus*, also from Mysore.

(v) An albino individual of an Indian Jackal, *Canis sp. incert.*

(vi) Two individuals of the Panda, or Red Cat-Bear, *Aelurus fulgens*, both over a year here.

(vii) Two Polar Bears, *Ursus maritimus*, which Mr. Hughes told me were brought to Mysore in 1906, and after seven years of life in the Tropics are in excellent condition.

There is a widespread and popular idea that animals found living in northern latitudes must necessarily like a low temperature, and also that bears, and especially polar bears, love cold weather. As far as my personal experience goes, there is no truth in this idea. These polar bears at Mysore have a large cage, with a tank of water and a very thick roof of thatch to ward off the rays of the midday sun, but I have seen other polar bears in far less pleasant quarters and yet looking fit and contented with

---

* Probably the *Loris lori lydekkerianus* described by Cabrera in 1908.

life in spite of a high temperature. This was in a travelling menagerie * in Genoa in 1903, and I was told that the bears were then doing their second summer in Italy and had been all the time in their travelling cage and had never had access to a bath, but were only from time to time washed down with a hose.

The important factor in keeping wild animals well and happy in captivity is not the size of their cage but the personal attention given to them by their keeper.

Mr. Hughes told me of a curious episode in which these bears took part. A short time ago a large python was found to have escaped from its cage in the Mysore garden during the night, and to have made its way into the quarters occupied by the polar bears who killed the python and ate part of it. Surely this must be the first record of polar bears making a meal off a snake !

(viii) A herd of thirteen albino Blackbuck, *Antilope cervicapra.* These white antelopes are regularly bred in Mysore and, I believe, nowhere else. A very remarkable fact Mr. Hughes told me is that the young bred from white parents are at first normally coloured, and as they become adult lose their pigmentation and become pure white.

(ix) A Markhor, *Capra falconeri.*

(x) At the time of my visit there was no specimen of the gaur, or Indian Bison, *Bos gavrus,* at Mysore. Mr. Hughes told me that only about a week before one had died from inflammation of the liver, and that another individual lived in this garden for eleven years, eventually dying of rinderpest.

(xi) Three Giraffes, *Giraffa camelopardalis (see* Pl. VI), a male and a female imported from Africa and a female born here. Their enclosure is surrounded by a fence of particularly light appearance, made out of metre gauge railway iron. Inside this fence there is a dry ditch about 4 feet (1·21 metres) deep and 10 feet

---

* J. Ehlbeck's Menagerie. There were five Polar Bears at Genoa, May 26, 1903.

(3·04 metres) wide. The giraffes apparently do not enter this ditch and try their strength against the fence beyond it. The vertical irons are built into a wall of masonry (which forms the outer edge of the ditch) and project 4 feet 10 inches (1·47 metres) above it. These verticals are placed about 23 feet (7·00 metres) apart, that is to say, the length of a metre gauge rail. The single horizontal bar is made of similar rails, fastened by fish-plates to the tops of the verticals. The horizontal irons are curved where necessary to follow the outline of the paddock. The outer surface of the wall, that is, towards the visitors, is covered with sloping turf.

(xii) The Nilgiri Blackbird, probably, *Merula simillina*.
(xiii) The Crested Bunting, *Melophus melanicterus*.
(xiv) Three Monáls, or Impeyan Pheasants, *Lophophorus refulgens*.
(xv) A male Somali Ostrich, *Struthio molybdophanes*.
(xvi) Three Indian Pythons, *Python molurus*, one of which was sitting on her eggs.

The pretty little striped Palm-Squirrels, *Sciurus palmarum*, live wild here. The Indian House-Crow, *Corvus splendens*, is very common, and also the Jungle-Crow, *Corvus macrorhynchus*, comes into the gardens. I actually saw a party of five or six of these black jungle-crows go in through the bars of the lion's cage and eat up the meat which had been placed there for the lion.

The Large Pied Wagtails, *Motacilla maderaspatensis*, looked very handsome walking on the lawns, and among the other wild things that seemed to find the Mysore Zoological Garden a pleasant home were many Vultures, *Neophron ginginianus*, and some very fine frogs, apparently *Rana tigrina*.

Besides the carnivora in the Zoological Garden, there are two very big male Mysore leopards kept in a cage in the courtyard of H.H. the Maharaja's Cow-House in the city of Mysore. I am told that to keep either leopards or chitas near the Cow-House is a local custom of great antiquity.

XIX.—NOTES AT PESHAWAR.

This Zoological Garden was started about 1909 and is situated in the "Shahi Bagh," or Imperial Garden, outside the city of Peshawar. The money for its upkeep is found by the Municipality, and the garden is open free to the public.

The Staff consists of :—

                1 Honorary Director. Capt. J. G. L. Ranking, Indian Army.
                1 Clerk, who is also the Resident Superintendent in Charge.
Gardens   1 Head gardener.
            20 Gardeners.
Menagerie 1 Salutri (native veterinary surgeon).
            2 Keepers.
            3 Assistant keepers.
            1 Bird-catcher and Aviary keeper.
            3 Night watchmen.
            1 Bhisti (water carrier).
            1 Grass-cutter, for animal's forage.
   Total 35.

Also ten Municipal sweepers assist in the Menagerie and gardens when required, making a total of forty-five available workers.

The Shahi Bagh abounds with wild birds. The bulbuls, mynas, bee-eaters, and doves are delightful, but the Crows, *Corvus splendens*, here, as in other zoological gardens in India, become a nuisance by worrying some of the larger animals, especially the Sámbar Deer, *Cervus unicolor*, by perching on their backs and pecking out bunches of their hair.

I visited the Peshawar garden on May 22 and 23, 1913, and found it one of the cleanest and best arranged menageries that I have ever had the good fortune to see.

The principal installations for animals are :—

(1) The Lion House (*see* Pl. VII).

A very good and carefully planned two-storied building. The lower floor is composed of a series of large outer cages (the fronts of which are shown in the photograph) and a corresponding series of smaller inner cages. The upper floor is a roomy service passage, from which the communication doors between the cages below are worked. These doors revolve on a vertical axis. The man above holding the handle has complete control over the door and can watch the movements of the animals through holes in the floor. The advantages of this system over the usual rising and falling door with a counterpoise balance are obvious. The animals are also fed from above, the keeper throwing them their food through a trap door, which gives the beasts good exercise as they spring upwards to catch the falling meat, and also affords a better spectacle for the visitors to watch than is provided in most zoological gardens, where the keepers thrust in the meat under the lowest bars of the cage front.

This house is to be completed by building a large open-air enclosure behind it, communicating with the inner series of cages. This exercising ground will be surrounded by high iron railings as in the Baroda, Bombay, and Calcutta menageries.

(2) The Aviary (*see* Pl. VIII).

A particularly well arranged building, with a service room in the centre surrounded by twelve cages, which are kept in beautiful order and are very nicely fitted up with growing plants, fresh branches of trees placed in barrels and pots, plots of grass and fountains of water.

(3) The Waterfowl Cage.

An extremely pretty and well planted enclosure with a pond. A delightful place both for birds to live in and for visitors to look at.

(4) The range of paddocks for ungulates.

(5) The Smaller Mammal House, where the wolf and hyaena live.

(6) The Otter Cage, with a brick-walled tank.

At Peshawar were two Black Celebes Apes. It may be mentioned that though Celebean monkeys are not very frequently seen in zoological gardens in Europe, I noted representatives of either *Cynopithecus niger* or various forms of what is commonly called *Macacus maurus* in the menageries at Bangalore, Baroda, Bombay, Calcutta, Karachi, Mysore, Peshawar, Rangoon, and Trivandrum.

Perhaps the most interesting of the animals living in the Peshawar collection was a Striped Hyaena, *Hyaena striata subsp.* (?), probably representing an undescribed race.

A very tame Frontier Wolf, *Canis lupus,* two Goral, *Cemas goral,* and a young Markhor, *Capra falconeri,* should also be specially mentioned.

As an example of the care which is bestowed on every individual animal in this garden, it may be recorded that the Monàl, or Impeyan Pheasant, *Lophophorus refulgens,* had been sent up to a hill station to avoid the hot season in Peshawar.

Two features of Peshawar, not connected with the zoological garden, but worth noticing, are :—

(1) That buffalo *bulls* are here used for riding and as beasts of burden.

(2) That the city pigeons are white, mottled, or particoloured, and not all Blue Rocks, resembling the wild *Columba livia* or *Columba intermedia,* as at Alwar and Jaipur in Rajputana.

XX.—NOTES AT RANGOON.

The Zoological Garden of Rangoon is called the Victoria Memorial Park ; it was founded in 1906. The site was given by the Government and the grounds laid out and buildings erected with funds subscribed by the public as a memorial to Queen Victoria. It is supported by contributions from the Government, from the Municipality, from the Port Commissioners, and by the gate-money, and is managed by a Board of Administrators through a committee of gentlemen who take special interest in the institution, one of the committee being an " Official Visitor " appointed by the Government.

The park adjoins the British Barracks and the Royal Lakes and is a very pretty piece of ground, with several small hills and nice ponds. The graceful golden pagoda, the Shwe Dagon, can be seen from many points in the grounds.

The area of the zoological garden was originally about 14 acres (5·66 hectares), is at present 35 acres (14·16 hectares), and is shortly to be extended and will then comprise about 80 acres (32·37 hectares).

The admission fee to the garden is 1 anna (1d., or 4 milliemes) per person.

An interesting pamphlet: "Notes on the Improvements of the Zoological Gardens, Rangoon," by Dr. R. M. Sen, the Superintendent, was published in 1910 under the authority of the Administrators of the Park, from page 24 of which we learn :—

" For the supply of foodstuff and other articles (such as kerosene oil, wire netting, nails, etc.) there is a contract renewed at the beginning of each official year by public tender at scheduled rates of prices plus a commission of certain percentage. Though not the cheapest, this appears to be the least troublesome process. On this, however, we cannot entirely depend. For certain articles, of which the necessity cannot be foreseen, direct purchase is the only course. There are others again, such as vegetables, fruits, fishes, etc., which it is desirable to purchase direct. If taken from the contractors, the choice is left with him whose interest it is to supply one of the many scheduled articles that will cost him the least. Direct purchase has the advantage of meeting urgent necessities, securing fresh stuff, and of selecting the desired articles, which can only be decided on by an inspection and enquiry in the market. The commission money paid last year approximated to 1,800 rupees, or 150 rupees a month."

I visited the Rangoon garden on May 5, 6, and 7, 1913, and am much indebted to Dr. A. Blake, to Mr. S. A. Christopher, to Mr. W. Shircore, Honorary Secretary, and to Dr. R. M. Sen, Superintendent, for giving me all the information I required.

The wild fauna of this beautiful garden add to its attractions; most noticeable were the brown squirrels, probably *Sciurus ferrugineus*, the Magpie Robins, *Copsychus saularis*, the Burmese Crows, *Corvus insolens*, the Myna Birds, *Acridotheres tristis* and *Sturnopastor superciliaris*, the Black-headed Chestnut Finches, *Munia atricapilla*, the pretty Brown-headed Sparrows, *Passer montanus*, the White-breasted Water-hens, *Amaurornis phoenicurus*, the Little Cormorants, *Phalacrocorax javanicus*,* the Eastern Cattle Egrets, *Bubulcus coromandus*, the little Chestnut Bitterns, *Ardetta cinnamonea*, and the very brilliantly coloured Lizards, probably *Calotes versicolor*, and, as in most gardens in India, the voices of the Crimson-breasted Barbet or Coppersmith, *Xantholaema haematocephala*, and of the Koel, *Eudynamis honorata*, could be very frequently heard.

---

* These cormorants are a great feature of Rangoon. I saw many flying over the main streets, and in the evening, about sunset, flocks of many hundred cormorants flew low along the river front, associated with the cormorants being some scores of eastern cattle egrets.

STAFF.

The Staff of the Rangoon Zoological Garden is as follows :—

Superior Staff	1 Superintendent : provided with a house in the garden.	
	1 Clerk (part time only).	
Gardens	1 Head gardener.	All Hindoos from Madras.
	1 Second gardener.	
	19 Gardeners.	
Menagerie	1 Store-keeper.	
	1 Head keeper.	All natives of India. The keepers are mostly Chittagong Mohammedans. I was surprised to find that no Burmese were employed here.
	6 Keepers.	
	9 Under keepers.	
	2 Gate keepers.	
	2 Peons (Messengers).	
	3 Day watchmen.	
	3 Night watchmen.	
	2 Carters.	
	1 Sweeper.	

Total 53

The most noticeable building in the Victoria Memorial Park is the Monkey House, originally the Phayre Museum, which stood in the town of Rangoon between Commissioners Road and Canal Street. The site, however, being required for other purposes, the building was taken down and eventually re-erected in the Zoological Garden and adapted to its present use. The contents of the Phayre Museum are said to be stored somewhere in Rangoon, but are not available for inspection.

The collection in the Zoological Garden is rich in Monkeys of the genus *Macacus*. Besides the species usually to be seen in menageries, there were a Himalayan Monkey, *Macacus assamensis*, said to be from beyond Bhamo, three Brown Stump-tailed Monkeys, *Macacus arctoides*, from Mergui, and three Burmese Pig-tailed Monkeys, *Macacus leoninus*.

Other inhabitants of the menagerie worthy of special notice are :—

(1) Three Slow Lorises, *Nycticebus. tardigradus subsp. incert.*

(2) Three Burmese Ferret-Badgers, *Helictis personata.*

(3) A Panda, or Red Cat-Bear, *Aelurus fulgens,* now two years here and fed almost entirely on bread and milk, its only other sustenance being a very little fruit.

(4) A large and beautiful Grey Flying Squirrel, *Pteromys sp.,* from Pegu.

(5) Four (or more) Bay Bamboo-Rats, *Rhizomys badius.*

(6) A male Asiatic Two-horned Rhinoceros, *Rhinoceros sumatrensis,* received from the Straits Settlements in 1909, and still tame enough to allow his keeper to sit on his back (*see* Pl. X).

(7) Three Malay Tapirs, *Tapirus indicus* (*see* Pl. XI); these animals breed regularly here and the young ones thrive.

(8) A young female Gaur, or Indian Bison, from Burma, so presumably representing *Bos gaurus readi* described by Mr. Lydekker in 1903.

(9) Three Gayal or Mithan, *Bos frontalis.*

(10) Five Tsaing, *Bos banteng birmanicus* * : three bulls and two cows. A young bull that has not yet assumed the dark colour that these animals get when old is figured on Plate XII.

(11) A male Anoa, *Bos depressicornis.* This animal is unfortunately blind in one eye, the injury having been caused by that frequent source of trouble in menageries—a Nilgai, *Boselaphys tragocamelus.*

(12) A nice herd of twelve Thameng, or Burmese Brow-antlered Deer, *Cervus eldi.*

The Emus, a Ceram Cassowary, a Yellow-necked Cassowary, and a gigantic land tortoise should also be mentioned.

(*) *Bus sondaicus* of Blanford, "Fauna of British India," Mammalia, 1891, page 489.

XXI.—NOTES AT TRIVANDRUM.

The Trivandrum Museum and Public Gardens, of which the menagerie forms part, were founded in 1859. They are a Government institution, included in the Department of Science and Art of the Travancore Government. A sketch of the origin and progress of these combined institutions has been written by Mr. H. S. Ferguson, the Director under whose care they became famous throughout the zoological world, and published in the Report on the Trivandrum Museum for M.E. 1075 (1899–1900 A.D.). After Mr. Ferguson's retirement the duties of Director were for some years (1904 to about 1910) undertaken by Colonel F. W. Dawson, then commanding the Nayer Brigade, whose headquarters are at Trivandrum. The Diwan Sahib P. Rajagopala Chari, C.I.E., the present Prime Minister, takes great interest in this Department, and Mr. A. J. Vieyra, Chief Secretary of the Travancore Government, now combines with his other duties that of Director of the Museum and Gardens.

This most interesting institution is open free to the public daily. I visited it on April 23, 24, 25, and 26, 1913, and Mr. Vieyra and his staff were most kind in giving me all the information I required. Mr. Thomas Alveyn is Chief Clerk, R. Shunkaranarayana Pillay is Chief Taxidermist, C. Subraya Mudaliar is Draftsman and Modeller, and K. G. Kesava Pillay is Head keeper.

The total area of the grounds is about $50\frac{1}{2}$ acres (20·43 hectares). The garden is a remarkably pretty one, all up and downhill, with deep valleys and steep slopes. The highest spot is occupied by the Museum, from which flights of brick steps lead down from tier to tier to a pretty lake. The botanical collection is a very rich one; my attention was specially called to the Sealing Wax Palm, *Cercostacis arunda*, and to the large series of beautiful orchids. The houses and paddocks for the zoological collection are scattered about the grounds in such a way as to add interest to the gardens *as gardens*, and yet not to spoil the effects of green lawns and rich tropical vegetation.

Little striped squirrels abound here, and also many wild birds (I noted forty-four different species in and about Trivandrum during the few days I was there), as well as very gaily coloured lizards, including *Calotes versicolor* and a red-tailed skink.

A watchman, armed with bow and pebbles, patrols the garden to warn off the Crows, *Corvus splendens*, from the animal's food.

The principal buildings in the Trivandrum Public Gardens are :—

(1) The Library.

This contains, besides a large scientific library, the office of the institution, a collection of geological specimens and minerals, some valuable archaeological exhibits, and a collection of pictures, one of which, a fine painting of the Durbar held by H.H. Marthanda Varma at Theketheruvoo Trivandrum in 1851, is of zoological interest, as in it a live giraffe is seen being paraded with the state elephants. I am told that there are records of two giraffes having been in the possession of the Maharajas of Travancore about that period.

(2) The Museum (*see* p. 94).
(3) The Reptile House.

This consists of one great hall, with a door in the middle of each side and with large windows to allow of as much ventilation as possible. The cases for reptiles form a quadrangle on the floor of this hall, the space between the cases and the walls forming a passage for the visitors, and the space enclosed by the inner sides of the cases forming a convenient service yard for the staff. Additional cages occupy the four corners of the hall. The special feature aimed at in the construction of this house was that the reptiles should be always in the shade, so the whole hall is covered by an overhanging roof, designed so that no direct sunshine can ever fall on the cages.

The crocodiles, which love to bask in the sunshine, are not kept in this house, but have special tanks allotted to them in the open air. The large python is also kept in an outdoor but roofed cage.

(4) The Lion House.
(5) The new Monkey House, now in course of construction.

STAFF.

The Staff of Trivandrum Public Gardens is as follows:—

	1 Director, unpaid as such, but receiving 100 rupees a *month* allowance.

Office.   5 Clerks.

Museum.   1 Head Taxidermist, at sixty rupees a month.

3 Assistant Taxidermists.

1 Draftsman and Modeller.

1 Assistant Draftsman.

1 Collector.

1 Assistant Collector.

5 Peons (as messengers, day watchmen, etc.).

Gardens.   1 Curator.

1 Assistant Curator.

1 Garden Assistant.

1 Maistry.

2 Peons (as messengers day watchmen, etc.).

10 Night watchmen : five of whom must always be on duty.

31 Gardeners.

Menagerie.   1 Veterinary surgeon, unpaid as such, but receiving 100 rupees a *year* allowance.

1 Head keeper, at thirty rupees a month.

3 Peons (as messengers, day watchmen, etc.).

1 Smith.

1 Bellows boy.

1 Carpenter.

1 Mason.

12 Keepers.

Total 87 ; or 85 without including the partially paid Director and Veterinary Surgeon.

Some women and girls are also employed in keeping the gardens tidy, and in addition temporary labourers are taken on as required.

Museum.

The Trivandrum Museum, though now called the "Napier Museum," was originally started by General Cullen. It is an ornate building, well lighted and well ventilated. The exhibition hall is about 60 yards (54·86 metres) in length. This contains :—

(1) An index collection of zoology.

(2) A collection of Travancore animals, of which the vertebrates are really very complete and well labelled.

(3) A collection of horns and antlers of big game from various parts of India and Central Asia : this is the " Sterndale Collection," of which a catalogue (eighteen pages) by Mr. R. A. Sterndale was published in 1889.

(4) An ethnographical collection of great interest.

The glass cases in which the stuffed mammals are exhibited were originally made as air-tight as possible, but the condition of the specimens has, I am told, been much improved by in each case removing one piece of glass and covering the opening with galvanized wire netting of half-inch mesh, so as to allow of a free circulation of air.

Special attention should be called to the following specimens in the Trivandrum Museum :—

(i) The collection of local Cetacea, including skeletons and very excellent models, coloured from life, of :—

Little Indian Porpoise, *Phocaena phocaenoides* (2 specimens).

Catalanian Dolphin, *Tursiops catalania*.

Col. Dawson's Dolphin, *Tursiops dawsoni*.

Speckled Dolphin, *Sotalia lentiginosa* (2 specimens).

The Dolphin, **Delphinus delphis**.

False Killer-whale, *Pseudorca crassidens*, about 16 feet (4·87 metres) long; and placed in the west porch of the Museum are two jaw-bones of a great Indian Fin-whale, *Balaenoptera indica*, from near Cape Comorin, March 16, 1904. The length of each bone is about 18 feet 3 inches (5·56 metres).

(ii) Two skulls with fine horns of the Gaur, or Indian Bison, *Bos gaurus*, from animals shot by Mr. Munro, near Devicolum.

(iii) A case of coloured plaster casts of the local poisonous snakes.

(iv) A specimen of the King-Cobra, or Hamadryad, *Naia bungarus*, 14 feet 4 inches (4·36 metres) in length, said to be a record for Travancore. ·

(v) Very good coloured plaster casts of the Frog, *Rana hexadactyla*, and of the Toad, *Bufo parietalis*.

And many nice plaster casts, coloured from life, of local fish, among the most striking being : —

(vi) *Zygaena tudes*, a species of Hammerheaded Shark.

(vii) A specimen of the harmless Giant Shark, *Rhinodon typicus*, from Trivandrum, February, 1909, 13 feet 7 inches (4·14 metres) in length. Another fish of this species was once beached at Trivandrum which was 29 feet (8·83 metres) long. "This shark has been said to exceed fifty feet in length, and some authors even assert seventy." (*Vide* Day, "Fauna of British Indies," Fishes, I, 1889, p. 29.) *

(viii) A Ray, *Rhinobatus granulatus*.

(ix) *Rhynchobatus ancylostomus*, the curious Mud-Skate. Two models, one of a dark adult, one of a pink youngster with conspicuous black marks.

(x) *Rhynchobatus djeddensis*, the Djedda Mud-Skate. Two models, an adult and a young specimen.

Menagerie.

The collection of monkeys included a large male Mias or Orang-Utan, *Simia satyrus*, and a specimen of the Nilgiri Langur, *Semnopithecus johni*. The Trivandrum Zoological Garden holds some good records of longevity ; thus a Bonnet

---

* A Shark of this species, forty-five feet long, was captured on June 1, 1912, near Knight's Key, Florida. *Vide* "New York Zool. Soc. Bulletin," XVI, No. 60, Nov. 1913, page 1047.

Monkey, *Macacus sinicus*,* received December 29, 1896, died of a wound on May 1, 1909, thus living here thirteen years four months four days; and a "*Macacus fuscatus*," purchased February 7, 1895, was destroyed, having gone blind, December 2, 1912, thus living here seventeen years nine months twenty-five days.

Among the Carnivora attention may be called to three tigers, all from Travancore, and one caught within thirty miles of Trivandrum. † Two of these tigers were caught in pits which had been dug to trap wild elephants in. Black Leopards, *Felis pardus var. melas*, breed in this menagerie regularly, and I was shown a female born here March 14, 1899, *i.e.* fourteen years one month twelve days, and still alive.

Representatives of the family *Viverridae* seldom live over ten years in captivity, but at Trivandrum there were a Malabar Civet-Cat, *Viverra civettina*, received December 1, 1897, *i.e.* fifteen years four months twenty-five days, and still alive; an Indian Palm-Civet, *Paradoxurus niger*, received November 14, 1898, *i.e.* fourteen years five months twelve days, and still alive; a Bear-Cat, *Arctictis binturong*, received December 8, 1901, *i.e.* eleven years four months eighteen days, and still alive; and I was told that a Stripe-necked Mongoose, *Herpestes vitticollis*, lived here for twelve years ten months and eighteen days. It is a curious coincidence that these four long-lived animals should all be males (this I have on the authority of the Head Keeper), and that also among the *Arctoidea* the three oldest animals in the Trivandrum menagerie should also be of this sex. These are a male Otter, *Lutra vulgaris*, received September 11, 1900, *i.e.* twelve years five months twenty-five days, and still alive; a male Smooth Indian Otter, *Lutra macrodus*, received November 2, 1897, *i.e.* fifteen years five months twenty-four days, and still alive; and a male Sloth-Bear, *Melursus ursinus*,

---

* "*Simia sinica*, Linnaeus, Mantissa Plantarum, Appendix, p. 521 (1771) (*ex* Buffon, vol. xiv, p. 241, Bonnet Chinois.) This name was not given from any idea that it was a native of China, but on account of the resemblance of the arrangement of the hair of the scalp to a Chinese hat." W. H. Flower, "Catalogue of the Museum of the Royal College of Surgeons," Part II, (1884), page 33.

† I was told that only a few years ago a tiger made its way one night into the heart of the city of Trivandrum and was shot by a Nayer sentry on duty at the Maharaja's Palace.

received February 13, 1895, *i.e.* eighteen years two months thirteen days and still alive.

The most valuable animal living in-this collection is a female Gaur, or Indian Bison, *Bos gaurus*, caught when a small calf in an elephant pit in Travancore on February 5, 1908, and now a magnificent beast in really beautiful condition, and so powerful that she is kept behind a fence of solid iron bars 1½ inches (0·038 metre) in diameter, and lately when frightened by a passing motor car she actually *bent* one of these stout bars !

Trivandrum was the only Zoological Garden in India where I saw representatives of the order *Edentata*. These were an Indian Pangolin, *Manis pentadactyla*, and an American Ant-eater, *Tamandua tetradactyla*.

The "doyen" of the parrots is a Red-and-Blue Macaw, *Ara macao*, received September 3, 1894, *i.e.* eighteen years seven months twenty-three days here and still alive.

After Calcutta, Trivandrum has the best collection of birds of prey of any Indian Zoological Garden that I visited, including a beautiful Crested Serpent-Eagle, *Spilornis cheela*. The Brahminy Kite, *Haliastur indus*, is kept in captivity here so that the inhabitants of Trivandrum may be certain of knowing where to see one, as there is a custom in Travancore for a man to make a vow that he will eat no food till he has seen a Brahminy Kite.

The White-necked Stork, *Dissura episcopus*, which is represented in the collection, is, I am told, considered a sacred bird in the north of Travancore.

The Ceram Cassowary, *Casuarius galeatus*, does very well here; the three birds I saw had all been bred at Trivandrum, hatched on the following dates :—

May 2, 1903 ; May 3, 1903 ; May 22, 1904.

In a tank in the garden are six Indian Terrapins, *Nicoria trijuga*, all said to be over twenty years in the collection. The oldest was obtained October 31, 1890, *i.e.* twenty-two years five months twenty-six days and still alive.

Of the snakes the most remarkable individual is a Malay Python, *Python reticulatus*, purchased for 50 rupees (£ 3. 6s. 8d.) December 23, 1898, and now estimated to be 19 feet (5·79 metres) long.

An Indian Tigrine Frog, *Rana tigrina*, "caught in November, 1904, did not eat till July 6, 1905, when it showed signs of emaciation. .It was fed by force on fish, and ever since that has been getting on well on this diet, but rather shy in the presence of the attendant." Report, Trivandrum Museum, 1904–1909, page 7.

INDEX OF GROUPS OF ANIMALS MENTIONED IN REPORT.

I.N. 6595-1918-500 br.

PLATE I.

BIG BULL ELEPHANT BELONGING TO H.H. THE MAHARAJA OF ALWAR.